DATE DUE

FE 19 08		
MR 11 08		

DEMCO 38-296

Studying the dynamics of a large number of particles interacting through long-range forces, commonly referred to as the N-body problem, is a central aspect of many different branches of physics. In recent years, significant advances have been made in the development of fast N-body algorithms to deal efficiently with such complex problems. This book is the first to give a thorough introduction to these so-called tree methods, setting out the basic principles and giving practical examples of their use.

After a description of the key features of the hierarchical tree method, a variety of general N-body techniques are presented. Open boundary problems are then discussed, as well as the optimization of tree codes, periodic boundary problems, and the fast multipole method.

No prior specialist knowledge is assumed, and the techniques are illustrated throughout with reference to a broad range of applications. The book will be of great interest to graduate students and researchers working on the modeling of systems in astrophysics, plasma physics, nuclear and particle physics, condensed-matter physics, and materials science.

MANY-BODY TREE METHODS IN PHYSICS

MANY-BODY TREE METHODS
IN PHYSICS

SUSANNE PFALZNER

Max-Planck-Research Unit
"Dust in Starforming Regions,"
University of Jena

PAUL GIBBON

Max-Planck-Research Unit
"X-Ray Optics,"
University of Jena

CAMBRIDGE
UNIVERSITY PRESS

te of the University of Cambridge
on Street, Cambridge CB2 1RP
York, NY 10011-4211, USA
h, Melbourne 3166, Australia

© Cambridge University Press 1996

First published 1996

Printed in the United States of America

Library of Congress Cataloging-in-Publication Data
Pfalzner, Susanne.
Many-body tree methods in physics / Susanne Pfalzner, Paul Gibbon.
p. cm.
Includes bibliographical references and index.
ISBN 0-521-49564-4 (hardcover)
1. Many-body problem. 2. Algorithms. 3. Mathematical physics.
I. Gibbon, Paul. II. Title
QC174.17.P7P44 1996 95-36237
530.1'44 – dc20 CIP

A catalog record for this book is available from the British Library.

ISBN 0-521-49564-4 hardback

To our daughter, Theresa,
and our parents, Helga and Hans and Elin and David

Contents

Preface

The difficulty in writing a 'how-to' book on numerical methods is to find a form which is accessible to people from various scientific backgrounds. When we started this project, hierarchical N-body techniques were deemed to be 'too new' for a book. On the other hand, a few minutes browsing in the References will reveal that the scientific output arising from the original papers of Barnes and Hut (1986) and Greengard and Rohklin (1987) is impressive but largely confined to two or three specialist fields. To us, this suggests that it is about time these techniques became better known in other fields where N-body problems thrive, not least in our own field of computational plasma physics. This book is therefore an attempt to gather everything hierarchical under one roof, and then to indicate how and where tree methods might be used in the reader's own research field. Inevitably, this has resulted in something of a pot-pourri of techniques and applications, but we hope there is enough here to satisfy the beginners and connoisseurs alike.

1

Introduction

Classical systems that consist of many particles interacting through long-range forces have interested physicists for centuries. The equation of motion of a system with more than two particles does not have an analytical solution, and it is only since the advent of high-speed computers that the trajectories of many particles could be followed simultaneously in detail. Over the last 40 years, computer simulations of N-body systems have become an indispensable tool in all branches of physics. Mathematically the N-body problem is represented by the solution of N second-order differential equations of the form

$$m_i \frac{d^2 \mathbf{r}_i}{dt^2} = -\nabla_i V \qquad\qquad i = 1, 2, 3 \dots N, \qquad (1.1)$$

where $\mathbf{r}_i$ and m_i are the positions and masses of the ith particle. The knowledge of the positions and velocities as a function of time allows the global or macroscopic properties of the system to be calculated.

The potential V in Eq. 1.1 can include different kinds of interactions – those stemming from the forces the particles exert on each other and those of external fields V_{ex} like external electric or gravitational fields. For classical systems the general form of the potential can be

$$V = V_{short} + V_{long} + V_{ex}, \qquad (1.2)$$

where V_{short} is a rapidly decaying function of distance, like, for example, the Van der Waals potential in chemical physics, and V_{long} is a long-range potential like, for example, the Coulombic or gravitational potential. For a comparison of a typical short-range potential with a long-range potential see Fig. 1.1. The external field V_{ex} is a function which is usually independent of the number and relative position of the particles and is calculated separately for each particle, which leads to a computation time of the order $O(N)$. In the numerical evaluation of fields the cost of computing V_{short} is of the order $O(N)$ too, because

1

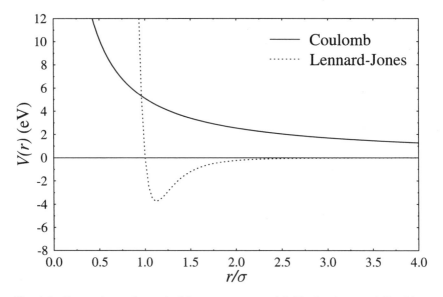

Fig. 1.1. Comparison of a typical long-range potential (Coulomb potential) with a short-range potential (Lennard–Jones potential).

the potential decays rapidly and each particle interacts significantly with only a small number of its nearest neighbours.

Although computers have made the simulation of this type of N-body problem feasible, those including long-range forces – Coulombic or gravitational – still present a challenge. The evaluation of V_{long}, if done directly, requires on the order of $O(N^2)$ operations, because this kind of potential decays slowly and the interaction of each pair of particles in the system has to be taken into account. Improvements in technique and computer speed have significantly increased the manageable simulation size, but the number of particles in such direct calculations is still too small for a variety of problems at present. This kind of computation is reasonable for a system involving a few hundred particles, but the costs increase so rapidly that simulations with a few thousands of particles are quite expensive and those with millions of particles are unattainable.

On the other hand, there are many examples in classical physics where models based on large-scale ensembles of particles interacting by long-range forces are very useful – astrophysics and plasma physics are two prominent examples. Several different approaches have been developed to reduce the burden of the long-range part of the calculation. Until recently, so-called particle-in-cell (PIC) methods have been regarded as the only effective way to simulate large systems of particles interacting through long-range forces. For a detailed review see Dawson (1983). The basic procedure of the PIC method is:

- A regular grid is laid out over the simulation region and the particles contribute their masses, charges, etc. to create a source density; this source density is interpolated at the grid points;
- the solution of the elliptical partial differential equation, usually obtained with the help of a fast Poisson solver, is used to calculate the potential values at the grid points;
- using these potential values the force is evaluated and interpolated to the particle positions.

So, by superimposing a grid of sample points, the potential field associated with the forces is calculated.

The total operation count for the PIC method is of the order $O(N + M \log M)$, where M is the number of mesh points and N is the number of particles in the system. Although the asymptotic computation cost is of the order $O(N \log N)$, in practice $M \ll N$ and, therefore, the numerical effort is observed to be proportional to N.

Due to this good computational efficiency, PIC codes are applied successfully to a variety of problems. However, there are three situations that PIC codes have difficulties dealing with:

- Strongly nonuniform particle distributions.
- Strongly correlated systems.
- Systems of complex geometry.

The first problem concerns the fact that the mesh in the standard particle-mesh schemes provides limited resolution. Due to the limitations of memory space in currently available computers, it may not be possible to use standard PIC methods to model the dynamics of a system with highly nonuniform source distributions, like, for example, in galaxies or in the cold dark matter scenario.

There have been attempts to overcome this disadvantage of the standard PIC method to obtain a better resolution. Villumsen (1989) has developed a PIC code which employs meshes of finer gridding in selected subregions of the system. Owing to the local improvement of the spatial resolution, these hierarchical meshes permit a more accurate modelling of regions of higher particle density. Where the poor resolution is of a dynamical nature – for example, due to shock waves – moving grids and adaptive grid refinement (Brackbill & Ruppel 1986) can be applied.

If the ratio of particles per cell is suitably large, local interactions are smoothed away in PIC simulations, leaving behind the collective or global behaviour of

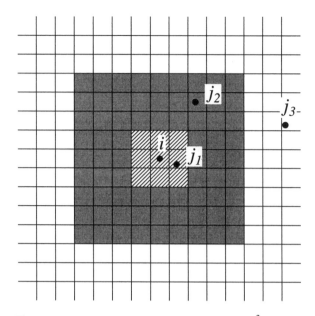

Fig. 1.2. Organisation of force contributions in P^3M code.

the system, and fluctuations due to the poor statistics are less of a problem. Although this smoothing effect is desirable in the context of, say, plasma kinetic theory, where the systems are usually considered uncorrelated, the PIC method is often not satisfactory where local corrections become important for the correct description of the physics. Many gravitational systems as well as high-density plasmas cannot be approximated as collisionless, and the two-body correlation function plays an important role.

A better method to describe the local correlations with PIC simulations is the so-called particle–particle particle–mesh technique (P^3M) (Hockney & Eastwood 1981). The idea in P^3M is to correct the far-field solution by including local forces by direct particle–particle interactions (see Fig. 1.2). This method seems to be an effective compromise between the possible number of particles and the spatial resolution in dynamic problems.

P^3M performs very well if the particles are more or less uniformly distributed in a rectangular region and relatively low precision is required. However, if the required precision is high or the particle distribution is clustered, this algorithm is sometimes not ideal. In these situations, inaccuracies can be introduced by matching long- and short-range forces (Bouchet & Hernquist 1988), and the computational effort tends to become excessive; more direct particle–particle interactions have to be taken into account due to the clustering. This limits the number of particles that can be simulated and the degree of nonlinearity that can be handled.

Another problem due to the grid of the PIC algorithm is that it imposes boundaries. Such boundaries might be rather artificial if the system is in 'free space' or has a complicated geometry. For systems without a real physical boundary, such as the collision of two galaxies, this usually means that one has to construct a grid much larger than the actual space occupied by the particles to avoid unphysical particle-boundary effects.

For strongly coupled systems it would be desirable to use a direct particle–particle force calculation, if the number of simulation particles was not so limited due to the N^2 scaling of the computation time. In the mid-1980s, several workers devised hierarchical schemes to exploit the fact that a particle interacts strongly with its near neighbours, but less detailed information is needed to describe its interaction with more distant particles. The first codes of this kind were developed independently by Appel (1985), Jernighan (1985), and Porter (1985). Although these codes had a nominal $N \log N$ scaling for the computation time, they used neighbour lists and data structures which tended to become tangled, thus introducing errors due to unphysical groupings of particles. This problem was overcome by Barnes and Hut (1986), who used a tree structure rebuilt from scratch at each timestep, ensuring that particle groupings were systematically updated. An appealing feature of the Barnes–Hut scheme is that the $N \log N$ scaling can be rigorously proven.

In Chapter 2, the so-called hierarchical 'tree' algorithm will be described in its basic form. It will be shown how this special means to divide space is used to construct a hierarchical data structure. This tree structure provides a systematic way of determining the degree of 'closeness' between two different particles without explicitly calculating the distance between each particle pair. The net result is to reduce the computational effort of the force calculation to $O(N \log N)$. The force on an individual particle from other particles close by is, on average, evaluated by direct particle–particle interaction, whereas the force due to more distant particles is included as a particle-cluster contribution. To obtain a better accuracy, the multipole moments of the cluster can be included. For dynamical systems, this process of building a tree structure and using it for the force calculation is repeated at each step.

A fairly sizeable literature already exists on tree codes, including a number of works comparing their performance against standard PP codes – notably Hernquist (1987). It is beyond the scope of this book to provide an exhaustive list of 'validity' criteria for hierarchical algorithms, but a few of the more important tests, such as energy and momentum conservation, will be discussed at the end of Chapter 2. Specific examples will be described in Chapters 3 and 6.

In Chapter 3, applications of the tree algorithm will be described in astrophysics, beam transport, and nuclear physics. The examples discussed here have the common feature that the system evolves freely in space and no fixed boundaries are imposed. Open boundary problems are easiest to investigate with tree codes because one simply checks the size of the system after each timestep and adapts the simulation box accordingly so that all particles are included. Tree codes were first developed to study the collision of galaxies (Barnes & Hut 1986, Barnes & Hernquist 1993), where this adaptiveness of the simulation region is of great advantage. Some examples of such simulations which have led to a new understanding of galaxy dynamics and insight into the role of dark matter will be described.

The transport of particle beams in storage rings is of widespread importance for applications requiring a well-characterised and tightly focussed source of electrons or ions. Whether the beam is for etching grooves on an integrated circuit, or for generating X-rays in an inertial fusion hohlraum, it is essential to optimise the quality of the beam as it is accelerated and transported. After a brief introduction to elementary particle beam concepts, some applications are proposed where tree codes could challenge the near-monopoly on beam modelling currently enjoyed by PIC codes.

At one of the frontiers of fundamental physics is the field of heavy ion collisions. By smashing heavy ions together and observing the fragments which result from disintegrating nuclei, one can learn a great deal about how nuclear matter is held together. Through a combination of experiments and theoretical modelling, it is possible to deduce the properties not only of elemental nuclei, but also of more exotic objects and phenomena, such as neutron stars and the early stages of the universe. One promising approach for studying nuclear fragmentation is 'Quantum Molecular Dynamics' (QMD), which explicitly takes into account many-body correlations between nucleons. A detailed case study is presented here, showing how the tree algorithm can be integrated into existing QMD models.

The tree algorithm in Chapter 2 is described in its basic working form. In Chapter 4 it will be shown how to improve the performance of such a code, both through higher accuracy and standard optimisation techniques. A common problem in N-body simulations is that a relatively small number of particles undergo close encounters. Assuming that the necessary stability criteria are satisfied, these particles often determine the timestep on which the whole simulation has to be performed. One way to reduce the computation time employed in direct particle–particle calculations is to use an individual timestep for each particle. Due to the hierarchical structure this trick is not easy to combine with the tree algorithm, though a few implementations now exist. It will be shown that it is possible to have different timesteps for particles undergoing close encounters and particles belonging to the rest of the system, reducing the computation time significantly. Better accuracy of the particle path can be achieved by introducing higher order integration schemes. In addition to these software improvements, the performance of the tree code also depends on the computer hardware. Details will be given on how the tree algorithm can be restructured so that vectorisation and parallelisation architectures can be fully exploited.

There are many cases where it is not possible to simulate an entire physical system (e.g., a solid). On the other hand, it is reasonable to model a small part of it and take the rest into account by including periodic images of the simulation area. These so-called periodic boundary problems are treated in standard particle–particle codes either by the minimum image method (Metropolis et al. 1953) or the Ewald method (Ewald 1921). In the minimum image method, a box is formed around the individual particle, which is equal to the size of the simulation area. In the force calculation only the interactions with particles within this box are included. For the tree algorithm a difficulty arises because the cutting process tends to split the more distant groupings of particles. In Chapter 5 it will be shown how this method nevertheless can be adapted to tree codes. The minimum image method can be used only for weakly coupled systems; for strongly coupled systems the force of more distant particles has to be included too. In this case, the Ewald summation method is required, which includes an infinite number of periodic images by modifying the Coulomb potential – the so-called Ewald potential. Due to the fact that the tree algorithm deals with particle–pseudoparticle as well as particle–particle interactions, it is also necessary to include the higher moments of the multipole expansion of the Ewald potential (Pfalzner & Gibbon 1994).

During the last 15 years a lot of effort has been put into extending the applicability of the MD method to systems out of equilibrium. It will be shown how the tree method can be used for nonequilibrium simulations and where special care is needed to control thermodynamic quantities such as temperature. The properties of the periodic boundary system are often investigated, not only by dynamic methods (or molecular dynamics), but via static or Monte Carlo methods. A brief section is included to show how the tree algorithm can also be used to perform such calculations.

In Chapter 6 examples for the application of periodic tree codes will be described. One obvious candidate for periodic tree simulations is dense plasma. These relatively large systems interact by long-range Coulomb forces. Low density plasmas can be modelled successfully by hydrodynamic codes, which basically treat the plasma as ideal gas, or by PIC codes, which are more suitable for collisionless problems. As the density increases, collisions become important and have to be included in the calculation. Fokker–Planck codes model this by a collision term, but this procedure is usually somewhat artificial, and is valid only for small-angle scattering. Particle–particle codes have the advantage that they include collisions without any artificial assumptions. Static properties such as structure factors have been successfully calculated for very high densities by standard particle–particle codes, but dynamical properties and strongly nonlinear effects need a much larger number of simulation particles. It is precisely this sort of application for which periodic tree codes would make an ideal simulation tool.

In addition to these systems which interact purely by Coulomb forces the tree algorithm can be applied successfully to systems interacting via different force terms like in Eq. 1.2 where one of the terms is either gravitational or Coulombic. In these cases, the long-range term is usually by far the most time consuming and a speed-up by means of the tree algorithm can bring large gains. Different applications of tree codes to systems with a more complex structure of the potential are proposed here, among them ionic liquids, molten salts, and biological macromolecules like proteins.

Chapter 7 contains an introduction to the so-called Fast Multipole Method (FMM), which is, in some sense, based on a hierarchical tree code using a high-order multipole expansion. This kind of code is mathematically more complicated and has a higher computational effort than a standard tree code, but the computation time has in principle an $O(N)$ dependence. Formulations of the algorithm for both 2D and 3D problems will be outlined, based on works by Greengard and Rohklin (1987), and Schmidt and Lee (1991). An attempt is also made to compare the relative performance of the BH and FMM algorithms.

2

Basic Principles of the Hierarchical Tree Method

2.1 Tree Construction

We have seen in the preceding chapter that in grid-based codes the particles interact via some averaged density distribution. This enables one to calculate the influence of a number of particles represented by a cell on its neighbouring cells. Problems occur if the density contrast in the simulation becomes very large or the geometry of the problem is very complex.

So why does one bother with a grid at all and not just calculate the inter-particle forces? The answer is simply that the computational effort involved quite dramatically limits the number of particles that can be simulated. Particularly with $1/r$-type potentials, calculating each particle–particle interaction requires an unnecessary amount of work because the individual contributions of distant particles is small. On the other hand, gridless codes cannot distinguish between near-neighbours and more distant particles; each particle is given the same weighting.

Ideally, the calculation would be performed without a grid in the usual sense, but with some division of the physical space that maintains a relationship between each particle and its neighbours. The force could then be calculated by direct integration while combining increasingly large groups of particles at larger distances. Barnes and Hut (1986) observed that this works in the same way that humans interact with neighbouring individuals, more distant villages, and larger states and countries. A resident of Lower-Wobbleton, Kent, England, is unlikely to undertake a trip to Oberfriedrichsheim, Bavaria, Germany, for a beer and to catch up on the local gossip.

Independently, in the early 1980s, several workers attempted to implement this kind of hierarchical grouping in N-body codes (Appel 1985, Jernighan 1985, Porter 1985). Although these early hierarchical codes had a nominal $N \log N$ dependence of the computation time, additional errors were introduced that were hard to analyse because of the arbitrary structure of the tree. Complicated bookkeeping was required to 'reconnect' groups of near neighbours, with the result that the $N \log N$ scaling could only be conjectured.

Barnes and Hut (1986) introduced a scheme which avoided these complications. Its systematic division of the physical space has since become the basis of most hierarchical tree codes and their refinements. These codes are sometimes described as octagonal tree codes to distinguish them from so-called binary tree codes. Press (1986) and Benz (1988) have developed binary trees, based on nearest neighbour pairs, and Jernighan and Porter (1989) devised an integration method in which the particles and the nodes of the tree are the basic dynamical units. Although such binary trees might reflect the structure of the system more closely, the Barnes and Hut method is by far the most commonly used method due to its conceptual simplicity and straightforward tree construction. Therefore, we will restrict our discussion to 'oct-trees' throughout this book.

In their original work, Barnes and Hut (1986) began with an empty cubical cell that was big enough to contain the whole system of particles. Particles are placed one by one into this 'root' cell. If any two particles fall into the same cell (which happens as soon as the second particle has been loaded into the root cell), the cell is divided into daughter cells having exactly half the length, breadth, and width of their parent.

This means that in 3D, the system is split into eight pieces. If the two particles are still in the same daughter cell, this cell is recursively subdivided in the same way until the particles sit in different boxes. Then, the next particle is loaded and the same procedure starts again, except that the starting point is the level just above the root cell (because the latter has already been subdivided). When all N particles have been loaded, the system space will have been partitioned into a number of cubical cells of different sizes, with at most one particle per cell.

For numerical reasons which will be discussed later, most algorithms do not start with an empty box, but with the root cell containing all particles of the simulation (Hernquist 1988). As before, this cell is then divided into its eight daughter cells. For each cell one asks the question: How many particles are there in the cell – 0, 1, or > 1?

- If the cell is empty, this cell is ignored.

- If there is one particle in the cell, this is stored as a 'leaf' node in the tree structure.

- If there are more particles in a cell, this cell is stored as a 'twig' node and subdivided.

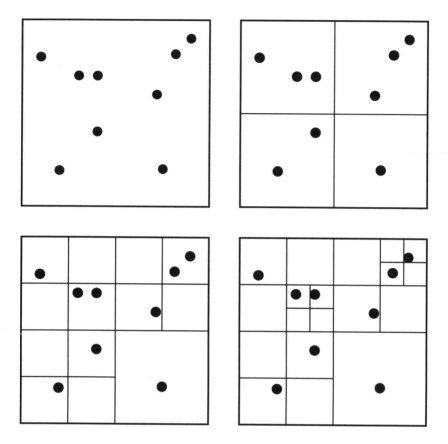

Fig. 2.1. Step-by-step division of space for a simple 2-D particle distribution.

The subdivision process continues until there are no cells with more than one particle left, which ultimately leads to an identical structure as the Barnes–Hut (BH) algorithm. Figure 2.1 shows the end-result of the space-division. For ease of illustration, a 2-dimensional example has been chosen; this means that one has always four daughter cells instead of eight.

Figure 2.2 shows the end-product of the division of space in three dimensions for the example of a star cluster (galaxy). Galaxies usually have a steep gradient in the particle distribution; they may be very dense at the centre relative to their outer reaches. While this is a source of difficulty for simulations with grid-based codes like PIC, it can be seen that the subdivision of space by the tree method *automatically* adjusts to the particle distribution.

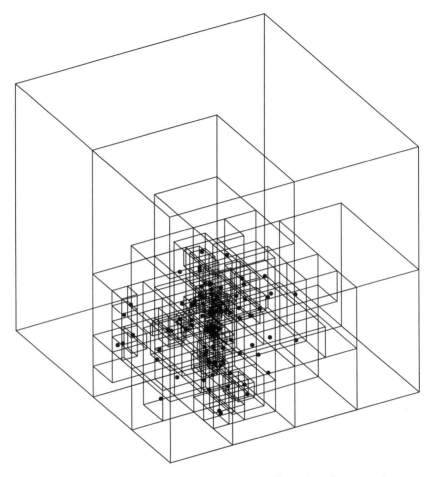

Fig. 2.2. Example of the division of space in three dimensions for a star cluster.

However, the division of space just described is not used like a grid, but as a bookkeeping structure for the tree. Figure 2.3 illustrates this relationship explicitly. The 'root' is the basic cell to start from, which contains all the particles of the simulation. At each division step the tree data structure is augmented with the next level in the hierarchy. Each node in the tree is associated with a cubic volume of space containing a given number of particles; empty cells are not stored.

The root cell is the biggest 'twig'. In our example in Fig. 2.3, the first division would lead to four nonempty cells:

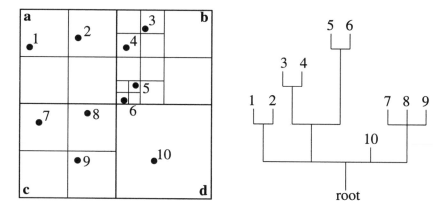

Fig. 2.3. Relationship between spatial division and the resulting tree structure.

- Cell **a** contains particles 1 and 2.
- Cell **b** contains particles 3, 4, 5, 6.
- Cell **c** contains particles 7, 8, 9.
- Cell **d** contains particle 10.

For the root, an identifier as twig as well as the number of nonempty daughter cells (here four daughters) have to be stored. Cell **d** contains only one particle, which means it is a leaf and needs no further processing. By contrast, cells **a**, **b**, and **c** are twigs, and are subdivided further.

For the leaf, an identifier as a leaf, a pointer to the parent cell (here the root), and the particle label have to be stored. This label will be the link to the physical quantities of the particle, like its position, mass, and charge. For a twig, the same information as for the root has to be stored *plus* a pointer to its parent cell. Having accumulated this information, the first level of division is finished. At the next level, the division of cell **a** leads to the creation of leaves 1 and 2 and the division of cell **c** leads to leaves 7, 8, and 9, whereas the division of cell **b** leads to two new twigs. Continuing this process, we finally end up with only leaves at the ends of the twigs. The whole data structure of the tree is thus in place.

Figure 2.4 shows a flowchart of the division process and the tree building. To actually convert this algorithm into a computer code, a number of arrays are necessary to define the tree structure with its cross references. In order to avoid being too language-specific, we will not show any explicit coding at this point,

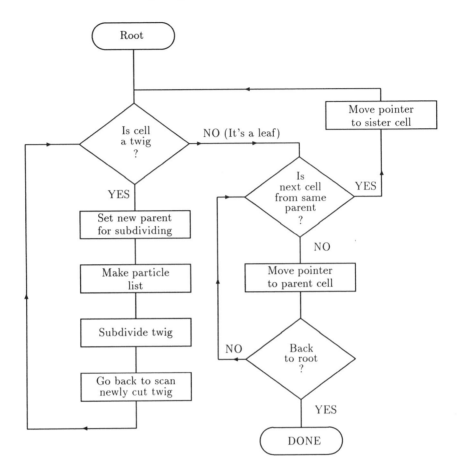

Fig. 2.4. Flowchart of tree building.

but it is perhaps instructive to consider the most important arrays for the example in Fig. 2.3. As the tree is built, each node is given a label itwig = (-1, -2, -3, ...) for the twigs and ileaf = (1, 2, 3, 4, ...) for the leaves. The root has a label 0. A pointer provides a link between the node number and the position in the array. Each node's parent and level of refinement is also stored. For each twig node, it is also useful to store the position of the 1st daughter node and the number of daughter nodes. For example, the command 'subdivide twig' in Fig. 2.4 actually refers to a subroutine containing the following operations:

SUBDIVIDE(cut-node,pointer)

```
iparent = cut-node
ipoi = pointer
ilev = level(iparent) + 1
1st_dau(iparent) = ipoi
```

count particles in subcells
n_dau(iparent) = # new daughter nodes

for each twig-node **do**:
```
    itwig = itwig − 1
    node(ipoi) = itwig
    parent(ipoi) = iparent
    level(itwig) = ilev
    ipoi = ipoi + 1
```
end do

for each leaf-node **do**:
```
    ileaf = ileaf + 1
    node(ipoi) = ileaf
    parent(ipoi) = iparent
    plabel(ileaf) = particle #
    level(ileaf) = ilev
```
    store position and moments of particle
```
    ipoi = ipoi + 1
```
end do

Before entering the SUBDIVIDE routine, the variable pointer is set at the end of the tree. On exit, it will point to the first node (twig or leaf) created in that subdivision. If this node is a twig, SUBDIVIDE is called again, otherwise we check for 'sister' nodes or descend back to the root. The arrays for the completed tree corresponding to Fig. 2.3 are shown in Table 2.1. Each of the intermediate horizontal lines in the table marks the beginning of a subdivision, that is: the point where SUBDIVIDE is called. The arrays 1st_dau and n_dau are filled in on the next level up, after the node has been subdivided.

A rough estimate of how many divisions are needed to reach a typical cell, starting from the root, can be obtained from the average size of a cell containing one or more particles. The average volume of such a cell is the volume of the

Table 2.1. *Tree arrays corresponding to Fig. 2.3*

pointer	level	node	parent	1st_dau	n_dau	plabel
1	0	0	0	2	4	–
2	1	-1	0	6	2	–
3	1	-2	0	8	2	–
4	1	-3	0	15	3	–
5	1	1	0	–	–	10
6	2	2	-1	–	–	1
7	2	3	-1	–	–	2
8	2	-4	-2	10	2	–
9	2	-5	-2	12	1	–
10	3	4	-4	–	–	3
11	3	5	-4	–	–	4
12	3	-6	-5	13	1	–
13	4	6	-6	–	–	6
14	4	7	-6	–	–	5
15	2	8	-3	–	–	7
16	2	9	-3	–	–	8
17	2	10	-3	–	–	9

root cell V divided by the number of simulation particles N. Moreover, the average length of a cell is a power of $V^{1/3}/2$. Therefore,

$$\left(\frac{1}{N}\right)^{1/3} = \left(\frac{1}{2}\right)^{x},$$

which means that the height x of the tree is of the order $\log_2 N^{1/3}$. This is equivalent to

$$\log_2 N^{1/3} = \frac{1}{3 \log 2} \log N \simeq \log N. \tag{2.1}$$

Starting from the root, an average of $O(\log N)$ divisions are necessary to reach a given leaf. The tree contains N leaves, therefore the time required to construct the tree is $O(N \log N)$.

As shown in the second loop of the SUBDIVIDE routine, an identifier (i.e., a numerical label) of the particle is stored for every leaf, which provides the link to the physical properties of the particle such as position, mass, and charge. Once the tree structure is completed, equivalents of these quantities, like the centre of mass, the sum of the masses, and the sum of the charges, still have to be calculated and stored for the twigs as well, because this information will be needed for the force calculation. This 'loading' of twig-nodes can be performed by propagating information down the tree from individual particles (leaves) towards the root. The tree structure can be used to find out which leaves and twigs belong to a given parent. The total mass is simply calculated by a sum over the masses of the daughter cells

$$m_{parent} = \sum_i m_i \, (daughters) \tag{2.2}$$

and the centre of mass by the sum

$$\mathbf{r}_{parent} = \frac{\sum_i m_i \mathbf{r}_i}{\sum_i m_i}. \tag{2.3}$$

Once these quantities have been defined for a twig-node, we can use them instead of the particle distribution inside it, that is, it can be regarded as a 'pseudoparticle'. Proceeding down the tree, the total mass and the centre of mass of the pseudoparticle can be used to define the pseudoparticles on the next lower level of division, as shown in the following MOMENTS routine.

MOMENTS

```
        ilevel = levmax − 1
        itwig = −ntwig


    for each level do:
        repeat for each node on same level:
            pointer = ipoint(itwig)
            nbuds = ndau(pointer)
            point1 = 1st_dau(pointer)
            zero moment sums:
            M(itwig) = 0
            r_com(itwig) = 0

            do i = 1,nbuds
                point_dau = point1 + i − 1
                inode = node(point_dau)
                sum mass and centre of mass:
                M(itwig) = M(itwig) + M(inode)
                r_com(itwig) = r_com(itwig) + r_c(inode)
            end do
```

$$\mathbf{r}_{com}(\texttt{itwig}) = \mathbf{r}_{com}(\texttt{itwig})/M(\texttt{itwig})$$

```
        itwig = itwig + 1
    until level done
    ilevel = ilevel + 1
end do
```

The variable definitions are the same as in Table 2.1. The simple loop over the daughter nodes is made possible because during the tree construction they are stored sequentially (see routine SUBDIVIDE). Eventually one reaches the root cell, the total mass of which is the same as the mass of the whole system. This is an excellent check to see that the tree structure is correct! Figure 2.5 illustrates graphically how the MOMENTS routine is used to calculate the total mass (denoted by the size of the symbol) and the centre of mass from the particles (black) through all levels for the pseudoparticles (grey).

2.2 Force Calculation

As we have seen, the tree structure needs a computational time of $O(N \log N)$ to build it. This structure is now used to perform the force calculation, which, as we will see, also requires a time $O(N \log N)$. The basic idea is to include contributions from near particles by a direct sum whereas the influence of remote particles is taken into account only by including larger cells, that is, pseudoparticles, representing many individual particles. This means that the advantages of the particle–particle technique are retained, but the computation time can be reduced significantly, especially for large numbers of particles (see Chapter 4). In contrast to grid-based codes, this time saving is achieved not by compromising the spatial resolution and/or imposing geometrical restrictions, but by introducing approximations into the calculation of the potential. This procedure is physically well motivated. For example, the dynamics of our solar system is insensitive to the detailed mass distribution of each planet. Furthermore, in all N-body simulations errors occur due to round-off, truncation, and discreteness effects, which makes it unreasonable to compute the potential field to extremely high precision. It is sufficient to require that the error in approximating the potential by the tree algorithm should be of the same order as these numerical errors. In dynamical applications one can often relax this requirement further because the errors introduced by finite timesteps in the integration scheme tend to dominate anyway.

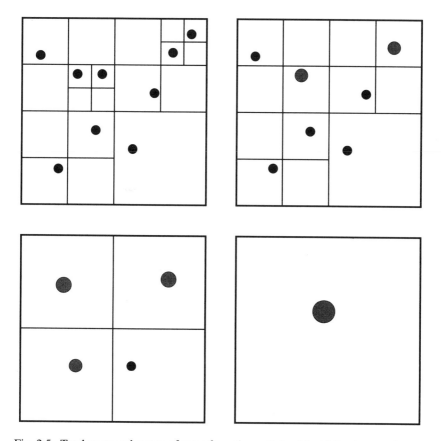

Fig. 2.5. Total mass and centre of mass from the particles (black) for the pseudoparticles (grey) at the different levels of the tree construction. The size of the symbol is proportional to the mass of the pseudoparticle.

The tree structure provides the means to distinguish between close particles and distant particles without actually calculating the distance between every particle. The force between near particles is calculated directly whereas more distant particles are grouped together to pseudoparticles. How do we actually decide when to group?

There is actually some flexibility in the way this is done, but we will start with the simplest method, introduced by Barnes and Hut (1986). Alternatives to the basic 's/d' criterion have recently been proposed by Salmon and Warren (1994), who demonstrate some potential problems with this choice by calculating 'worst-case' errors for various scenarios. A more detailed discussion of alternative criteria is deferred to Section 4.4.

For each particle the force calculation begins at the root of the tree. The 'size' of the current node (or twig), s, is compared with the distance from the particle, d. Figure 2.6 illustrates this comparison. If the relation

$$s/d \leq \theta \tag{2.4}$$

is fulfilled, where θ is a fixed tolerance parameter, then the internal structure of the pseudoparticle is ignored and its force contribution is added to the cumulative total for that particle. Otherwise, this node is resolved into its daughter nodes, each of which is recursively examined according to (2.4) and, if necessary, subdivided. Figure 2.7 shows a flowchart which demonstrates how the tree structure is used to calculate the forces. The node is subdivided by continuing the ascent through the tree until *either* the tolerance criterion is satisfied *or* a leaf-node is reached.

For nonzero θ the time required by the CPU (central processing unit) to compute the force on a single particle scales as $O(N \log N)$. Hernquist (1988) demonstrated this by using a simplified geometry. Consider a single particle at the centre of a homogeneous spherical particle distribution, with the number density n being

$$n = N \,/\, \left(\frac{4\pi}{3} R^3 \right),$$

where R is the radius of the whole system. Organising these particles in a hierarchical structure of cells, there will be an inner sphere where the force of each particle on the centre particle is directly calculated. The surrounding concentric shells will contain progressively larger pseudoparticles. Figure 2.8 illustrates this situation. The total number of interactions n_{int} can be estimated by

$$n_{int} = n_o + \sum_{shells} n^i_{sub},$$

where n_o is the number of direct interactions and n_{sub} is the number of subunits in each shell, which is roughly given by

$$n^i_{sub} \sim \frac{4\pi r_i^3 \theta}{\frac{4\pi}{3} r_i^3 \theta^3 / 8} = \frac{24}{\theta^2},$$

therefore

$$n_{int} \sim n_o + \frac{24}{\theta^2} n_{sh},$$

where n_{sh} is the total number of shells. The inner sphere, where only direct interactions are considered, has a radius of $r_1 = n^{-1/3}/\theta$, which leads to the

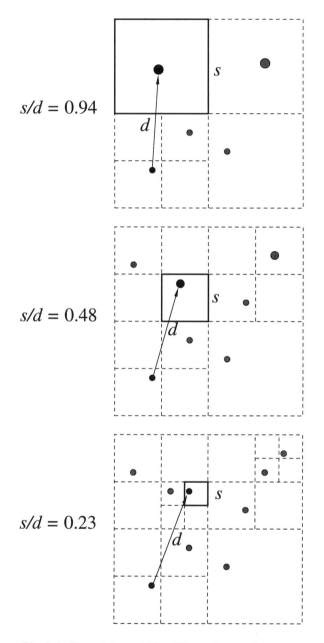

Fig. 2.6. The relation s/d for different levels of the tree.

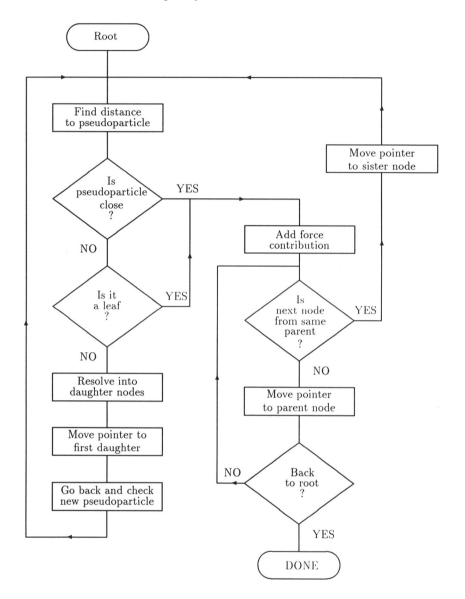

Fig. 2.7. Flowchart of the force calculation.

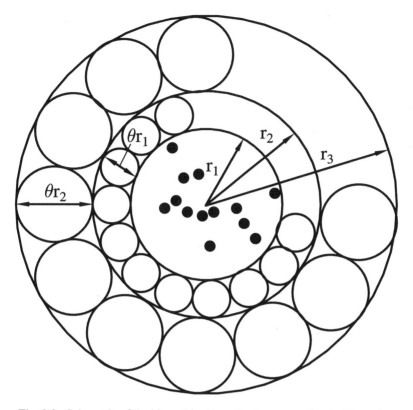

Fig. 2.8. Schematic of the hierarchical tree structure according to Hernquist.

number of interactions:

$$n_o = N \left(\frac{r_1}{R}\right)^3 = n\frac{4\pi}{3}r_1^3 = \frac{4\pi}{3}\frac{1}{\theta^3}.$$

Now the radii of the outer shells are $r_i = (1+\theta)^{i-1}r_1$, implying

$$\frac{r_n}{r_1} = (1+\theta)^{n_{sh}-1},$$

which gives the number of shells as

$$\begin{aligned}
n_{sh} &= 1 + \log_{(1+\theta)}\frac{r_n}{r_1} \\
&= 1 + \log_{(1+\theta)}\frac{R}{(1+\theta)r_1} \\
&= 1 + \log_{(1+\theta)}\left(\frac{1}{1+\theta}\frac{\theta}{(3N/4\pi)^{-1/3}}\right),
\end{aligned}$$

where we have used $R = r_n(1 + \theta)$ to get the last expression. Therefore, the total number of interactions is given by

$$n_{int} \sim \frac{24}{\theta^2} \frac{\log\left(\theta(3N/4\pi)^{1/3}\right)}{\log(1+\theta)} + \frac{4\pi}{3} \frac{1}{\theta^3}. \tag{2.5}$$

For large N, it follows that $\theta \geq (4\pi/3N)^{1/3}$; this means that the dominant behaviour for $\theta > 0$ is

$$n_{int} \sim \log N/\theta^2 \tag{2.6}$$

and the time required to calculate the force on a given particle is $O(\log N)$, which means the number of operations to compute the force on all N bodies will scale as $O(N \log N)$.

In contrast to the idealised picture in Fig. 2.8, the actual interaction list resembles one of the examples shown in Fig. 2.9. As this illustrates, the computation time not only is a function of the simulation size but also depends strongly on the choice of the tolerance parameter θ. The case $\theta = 0$ is equivalent to computing all particle–particle interactions, which is exact but rather time-consuming because the operation count is again $O(N^2)$. This is in fact slower than traditional PP because one has to build the tree and walk all the way up to the leaves during the force calculation. Choosing the other extreme, $\theta \to \infty$, would produce a very low spatial resolution with the force being calculated only between the individual particle and large pseudoparticles, which would be very quick but extremely inaccurate. This conflict in the choice of θ can be summarised as follows.

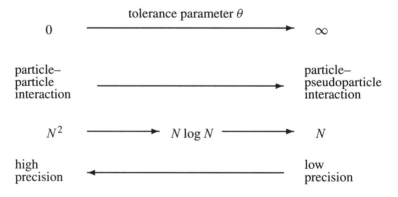

One would really like a rapid force calculation *and* high precision. In fact, a compromise of $\theta \sim 0.1 - 1.0$, depending on the problem, proves to be a practical choice. Fortunately, there is an additional way of improving the accuracy without the need of a higher resolution (smaller θ), which is to include

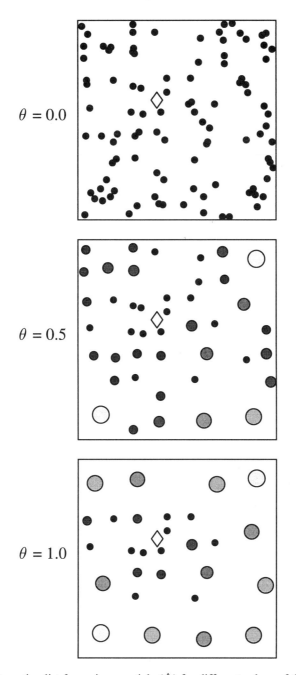

Fig. 2.9. Interaction list for a given particle ($\Diamond$) for different values of the tolerance parameter θ.

the multipole moments of the pseudoparticles, rather than treating them as point masses (or charges).

2.3 Multipole Expansion

In the analysis that follows in this section, we consider an N-body system which is characterised simply by the masses and positions of its N particles and has a $1/r$-potential. In the tree algorithm, distant particles are grouped to form a pseudoparticle with a mass equal to the sum of its constituent masses, and a position equal to the centre of mass of the group. This means a loss of information about the spatial distribution of the masses within the cell and leads to an error in the force calculation. One can recover the spatial information by performing a multipole expansion of the particle distribution in the cell. The force of the pseudoparticle on the individual particle is then given by

$$\mathbf{F}(\mathbf{R} - \mathbf{r}_i) \simeq \mathbf{F}(\mathbf{R}) \qquad \text{Monopole}$$

$$-\mathbf{r}_i \nabla \mathbf{F}(\mathbf{R}) \qquad \text{Dipole}$$

$$+\tfrac{1}{2}\mathbf{r}_i\mathbf{r}_i :: \nabla\nabla\mathbf{F}(\mathbf{R}) \quad \text{Quadrupole}$$

$$+\ldots,$$

where $\mathbf{R}$ is the vector from the particle to the centre of mass and $\mathbf{r}_i$ is the vector from the particle to an individual particle in the cell (see Fig. 2.10). If this multipole expansion is carried out to a high enough order, it will contain the total information of the particle distribution in the cell. There are more recently developed codes based on a 'Fast Multipole Method' (FMM) which exploit this fact. They use large pseudoparticles and perform the multipole expansion to a high order (typically 10–20) and represent a most elegant refinement of the tree code, which, depending on context and accuracy requirements, starts to become economic for $N \gtrsim 10^4$–10^5.

The maximum size of a pseudoparticle in a 3-dimensional code is $1/8$ of the total system volume. In practice, the size of the cell occupied by a pseudoparticle is determined by the choice of θ. For potentials falling off as $1/r$ or faster, higher moments contribute increasingly less to the force. In the range of $\theta \sim 0.1 - 1.0$ it turns out that including the dipole and quadrupole moments of the pseudoparticles gives acceptable accuracy in the force calculation for typical dynamical applications. Doing the tree construction by subdividing cubic cells fixes us to a Cartesian coordinate system and the multipole expansion is done in Cartesian coordinates too. Unfortunately, this is rather clumsy in comparison to the usual expansion in Legendre polynominals found in Jackson

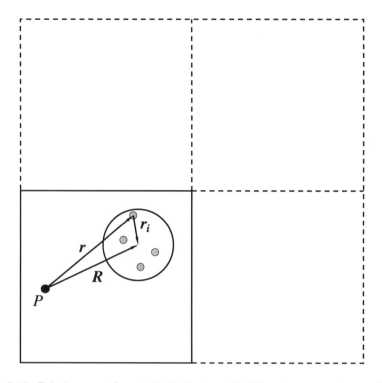

Fig. 2.10. **R** is the vector from the individual particle (P) to the centre of mass of the pseudoparticle (large circle), **r** is the vector to a single particle of this pseudoparticle, and $\mathbf{r}_i$ is the vector from a member of the pseudoparticle to the centre of mass.

(1975). Different formulations for computing higher order moments in 2- and 3-dimensional FMM codes are given in Chapter 7.

The potential at the origin Φ_o due to the pseudoparticle is the sum of the potentials Φ_i due to the particles in the cell,

$$\Phi(\mathbf{R}) = \sum_i \Phi_i(\mathbf{R} - \mathbf{r}_i),$$

where $\mathbf{r}_i$ is the vector from the particle to the centre of mass and the origin is, for simplicity, the individual particle on which the force of the pseudoparticle is calculated. Here we consider a $1/r$-potential, therefore

$$\Phi_i(\mathbf{R} - \mathbf{r}_i) = -\frac{m_i}{|\mathbf{R} - \mathbf{r}_i|}$$

$$= -\frac{m_i x_i}{\sqrt{(x - x_i)^2 + (y - y_i)^2 + (z - z_i)^2}}.$$

The multipole expansion of the potential up to the quadrupole moment is given by

$$\Phi(\mathbf{R}) = -\sum_i m_i \left[1 - \mathbf{r}_i \frac{\partial}{\partial \mathbf{r}} + \frac{1}{2} \mathbf{r}_i \mathbf{r}_i \frac{\partial}{\partial \mathbf{r}} \frac{\partial}{\partial \mathbf{r}} + \dots \right] \frac{1}{R}$$

$$= -\sum_i m_i \left[1 - x_i \frac{\partial}{\partial x} - y_i \frac{\partial}{\partial y} - z_i \frac{\partial}{\partial z} \right.$$

$$+ \frac{1}{2} x_i^2 \frac{\partial}{\partial x} \frac{\partial}{\partial x} + \frac{1}{2} y_i^2 \frac{\partial}{\partial y} \frac{\partial}{\partial y} + \frac{1}{2} z_i^2 \frac{\partial}{\partial z} \frac{\partial}{\partial z}$$

$$+ \frac{1}{2} x_i y_i \left(\frac{\partial}{\partial x} \frac{\partial}{\partial y} + \frac{\partial}{\partial y} \frac{\partial}{\partial x} \right)$$

$$+ \frac{1}{2} y_i z_i \left(\frac{\partial}{\partial y} \frac{\partial}{\partial z} + \frac{\partial}{\partial z} \frac{\partial}{\partial y} \right)$$

$$\left. + \frac{1}{2} x_i z_i \left(\frac{\partial}{\partial x} \frac{\partial}{\partial z} + \frac{\partial}{\partial z} \frac{\partial}{\partial x} \right) \right] \frac{1}{R}.$$

There are in principle three kinds of derivatives

$$\frac{\partial}{\partial x} \frac{1}{R} = -\frac{x}{R^3},$$

$$\frac{\partial}{\partial x} \frac{\partial}{\partial x} \frac{1}{R} = \frac{3x^2}{R^5} - \frac{1}{R^3},$$

$$\frac{\partial}{\partial x} \frac{\partial}{\partial y} \frac{1}{R} = \frac{3xy}{R^5},$$

and the other coordinates can be obtained in the same way. The multipole expansion of the pseudoparticle potential is then

$$\Phi(\mathbf{R}) = -\sum_i m_i \left[\frac{1}{R} + x_i \frac{x}{R^3} + y_i \frac{y}{R^3} + z_i \frac{z}{R^3} \right.$$

$$+ \frac{1}{2} x_i^2 \left(-\frac{1}{R^3} + \frac{3x^2}{R^5} \right) + \frac{1}{2} y_i^2 \left(-\frac{1}{R^3} + \frac{3y^2}{R^5} \right)$$

$$+ \frac{1}{2} z_i^2 \left(-\frac{1}{R^3} + \frac{3z^2}{R^5} \right)$$

$$\left. + x_i y_i \left(\frac{3xy}{R^5} \right) + y_i z_i \left(\frac{3yz}{R^5} \right) + x_i z_i \left(\frac{3xz}{R^5} \right) \right].$$

$$(2.7)$$

The force on the individual particle by the pseudoparticle is given as the derivative of the potential

$$\mathbf{F(R)} = -m_p \frac{\partial}{\partial \mathbf{r}} \sum_i \Phi_i. \tag{2.8}$$

Using Eq. 2.7 we can obtain the field by performing derivatives of the form:

Monopole (M):
$$\frac{\partial}{\partial x} \frac{1}{R} = -\frac{x}{R^3}.$$

Dipole (D):
$$\frac{\partial}{\partial x} \frac{x}{R^3} = \frac{1}{R^3} - \frac{3x^2}{R^5},$$

$$\frac{\partial}{\partial x} \frac{y}{R^3} = \frac{-3xy}{R^5}.$$

Quadrupole (Q):
$$\frac{\partial}{\partial x} \left(\frac{1}{R^3} - \frac{3x^2}{R^5} \right) = -\frac{9x}{R^5} + \frac{15x^3}{R^7},$$

$$\frac{\partial}{\partial x} \frac{-3xy}{R^5} = \frac{-3y}{R^5} + \frac{15x^2 y}{R^7},$$

$$\frac{\partial}{\partial x} \frac{-3yz}{R^5} = \frac{15xyz}{R^7},$$

$$\frac{\partial}{\partial x} \left(\frac{1}{R^3} - \frac{3y^2}{R^5} \right) = \frac{-3x}{R^5} + \frac{15xz^2}{R^7}.$$

The x-component of the force vector is then given by:

$$-F_x =$$

M:
$$\frac{x}{R^3} \sum_i m_i.$$

D:
$$-\left(\frac{1}{R^3} - \frac{3x^2}{R^5} \right) \cdot \sum_i m_i x_i + \frac{3xy}{R^5} \cdot \sum_i m_i y_i + \frac{3xz}{R^5} \cdot \sum_i m_i z_i.$$

Q:
$$+\left(\frac{15x^3}{R^7} - \frac{9x}{R^5} \right) \cdot \frac{1}{2} \sum_i m_i x_i^2 + \left(\frac{15xy^2}{R^7} - \frac{3x}{R^5} \right) \cdot \frac{1}{2} \sum_i m_i y_i^2$$

$$+\left(\frac{15xz^2}{R^7} - \frac{3x}{R^5} \right) \cdot \frac{1}{2} \sum_i m_i z_i^2 + \left(\frac{15x^2 y}{R^7} - \frac{3y}{R^5} \right) \cdot \sum_i m_i x_i y_i$$

$$+\left(\frac{15x^2 z}{R^7} - \frac{3z}{R^5} \right) \cdot \sum_i m_i x_i z_i + \left(\frac{15xyz}{R^7} \right) \cdot \sum_i m_i y_i z_i.$$

Equivalent expressions for the y- and z-components of the force can be easily obtained by cyclic rotation. Appendix 1 shows the multipole expansion for the somewhat simpler 2-dimensional case.

To keep the computational effort to a minimum it would be useful to utilize the information stored for the daughter nodes to obtain information about the parent nodes. The force expression contains sums such as: $M = \sum_i m_i$, the dipole moment $\mathbf{D} = \sum_i m_i \mathbf{r}_i$, and the quadrupole moments $Q_{xx} = \sum_i m_i x_i^2$, $Q_{xy} = \sum_i m_i x_i y_i$, and so forth. In general, the multipole moments depend on the choice of the origin of the coordinate system, but the value of the first non-vanishing moment is independent of this choice (Jackson 1975). This means that the monopole moment is independent of this choice, and therefore the total mass of the parent cell can simply be obtained by summing over the masses of the daughter cells.

The calculation of the dipole and quadrupole moments of the parent cell from the moments of the daughter cell is a bit more difficult because the moments depend on the choice of the origin. Initially, the moments are calculated relative to the centre of mass, defined by

$$\mathbf{r}_{cm} = \frac{\sum_i m_i \mathbf{r}_i}{\sum_i m_i}.$$

As Fig. 2.11 illustrates, the centres of mass are different for the daughter cells and the parent cell. The difference in the x-component is given by

$$x_{sd} = x_{cm}(daughter) - x_{cm}(parent),$$

where d indicates the different daughters. For example, for daughter 1 in Fig. 2.11, we have:

$$x_{s1} = X_1 - X_0.$$

This means that the moments of the daughter cell have to be calculated relative to a new origin, which is equivalent to a shift by x_{sd}. Each individual x_i in the sum of one daughter is shifted by the same vector x_{sd}, meaning: $x_i(\text{new}) = x_i - x_{sd}$. The dipole moment relative to the new origin can be obtained from the dipole moment relative to the original origin by making this substitution:

$$
\begin{aligned}
D_x^{daughter}(\text{new}) &= \sum_i m_i (x_i - x_{sd}) \\
&= \sum_i m_i x_i - \sum_i m_i x_{sd} \\
&= \sum_i m_i x_i - x_{sd} \sum_i m_i.
\end{aligned}
$$

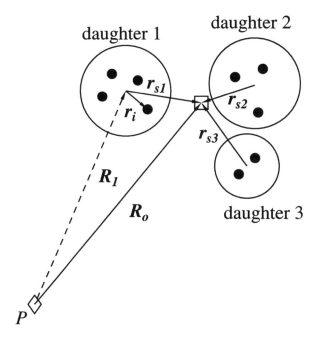

Fig. 2.11. Origin shift for the multipole calculation: the circles symbolize the pseudoparticles; $\mathbf{r}_i$ is the vector from a member of one of the pseudoparticles to the centre of mass of this pseudoparticle and $\mathbf{r}_{si}$ is the shifting vector to the new origin ($\square$), which is the centre of mass of the daughters (pseudoparticles).

The same procedure applies for the quadrupole moments, in which case we have:

$$
\begin{aligned}
Q_{xx}^{daughter}(\text{new}) &= \sum_i m_i (x_i - x_{sd})^2 \\
&= \sum_i m_i x_i^2 - 2x_{sd} \sum_i m_i x_i + x_{sd}^2 \sum_i m_i,
\end{aligned}
$$

and

$$
\begin{aligned}
Q_{xy}^{daughter}(\text{new}) &= \sum_i m_i (x_i - x_{sd})(y_i - y_{sd}) \\
&= \sum_i m_i x_i y_i - x_{sd} \sum_i m_i y_i \\
&\quad - y_{sd} \sum_i m_i x_i + x_{sd} y_{sd} \sum_i m_i.
\end{aligned}
$$

This means that all moments of the parent can be obtained from the knowledge of the daughters' moments and the shifting vectors $\mathbf{r}_{sd}$. The dipole moment of the parent cell is therefore:

$$D_x^{parent} = \sum_d \left(\sum_i m_i x_i - x_{sd} \sum_i m_i \right)$$

$$= \sum_d \left(D_x^d - x_{sd} M^d \right). \qquad (2.9)$$

Similarly, the quadrupole moments of the parent cell can be found from:

$$Q_{xx}^{parent} = \sum_d \left(\sum_i m_i x_i^2 - 2 x_{sd} \sum_i m_i x_i + x_{sd}^2 \sum_i m_i \right)$$

$$= \sum_d \left(Q_{xx}^d - 2 x_{sd} D_x^d + x_{sd}^2 M^d \right), \qquad (2.10)$$

$$Q_{xy}^{parent} = \sum_d \left(\sum_i m_i x_i y_i - x_{sd} \sum_i m_i y_i \right.$$

$$\left. + y_{sd} \sum_i m_i x_i + x_{sd} y_{sd} \sum_i m_i \right)$$

$$= \sum_d \left(Q_{xy}^d - x_{sd} D_y^d - y_{sd} D_x^d + x_{sd} y_{sd} M^d \right). \qquad (2.11)$$

How is this implemented in the code? Starting at the highest level of the tree, the sums $\sum_i m_i$, $\sum_i m_i x_i$, $\sum_i m_i x_i^2$, and $\sum_i m_i x_i y_i$ and the equivalent ones for the other space directions are calculated and stored for every pseudoparticle (twig node). The multipole moments for the next level down (i.e., for the parent nodes) are then evaluated *at their respective centres of mass* using the shifting formulae (2.9)–(2.11). In practice, this means adding another loop over the daughter nodes to the MOMENTS routine described earlier.

```
do i = 1, nbuds
    point_dau = point1 + i - 1
    inode = node (point_dau)

    shift vector
    rs = rcom(itwig) - rcom (inode)
```

sum moments

$\mathbf{D}$(itwig) = $\mathbf{D}$(itwig)
$\qquad$ + $\mathbf{D}$(inode) $- \mathbf{r}_s \times Q$(inode)

... etc.

end do

This procedure is continued until the root node is reached, which will then contain a (second-order) expansion for the whole system. The moments are later used to evaluate the dipole and quadrupole corrections in the force calculation.

For problems in which the forces are gravitational, the dipole moments vanish relative to the centre of mass, so that we have only monopole and quadrupole contributions to the force calculation. We will see later that this does not happen for electrostatic forces. Table 2.2 shows how including the moment terms in the force calculation improves the accuracy. Or, seen from a different point of view, the same level of accuracy in the force computation is achieved more efficiently if quadrupole moments are included rather than using a smaller θ in the monopole version. This is due to the fact that, as (2.5) shows, the number of interactions varies as θ^{-3}, and so increases rapidly with decreasing θ.

The error introduced into the force calculation as a result of the truncated multipole expansion increases monotonically with θ. For open systems the error relative to a direct particle–particle calculation is typically $\lesssim 0.1\%$ for $\theta = 1$. It increases rapidly for larger θ, which suggests that $\theta = 1$ may be a practical upper limit for open systems.

A factor of $\sim 1/\theta$ improvement of the relative force error $\Delta F/F$ is expected for each new added term in the expansion (McMillan & Aarseth 1993). However, it would be wrong to conclude that adding higher order multipole moments is always more effective than choosing a smaller θ. The evaluation time of the higher moments (2^l) of the multipole expansion has an $O(l^2)$ dependence. So, at some point, improving the accuracy by reducing θ costs less than including higher moments (Makino 1990a).

Table 2.2. *Force on a randomly chosen particle from an ion beam simulation*

N	PP code	Monopole tree		Dipole tree	
		$\Theta = 1.0$	$\Theta = 0.3$	$\Theta = 1.0$	$\Theta = 0.3$
100	0.5284	0.5141	0.5273	0.5271	0.5287
500	0.6911	0.7152	0.6932	0.6959	0.6912
1,500	0.8951	0.9139	0.8958	0.8965	0.8951

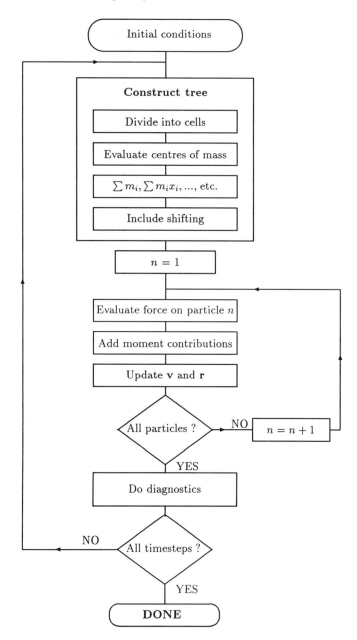

Fig. 2.12. Flowchart of a hierarchical tree code for a dynamic N-body problem.

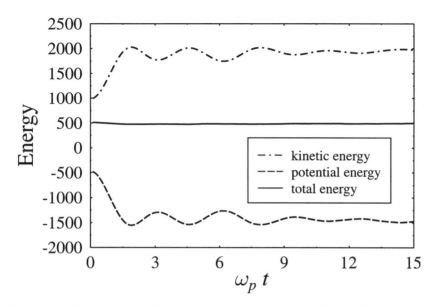

Fig. 2.13. Energy conservation for a plasma system approaching equilibrium over 10,000 timesteps.

2.4 Dynamics

Up to now we have considered only a single timestep of the dynamical system. In contrast to first hierarchical codes by Appel (1985), Jernighan (1985), and Porter (1985), recent codes mainly reconstruct the tree ab initio every timestep. This sounds like a lot of computational effort, but in fact only a small fraction of time is needed in generating the tree structure, which is usually of the order of a few percent for $\theta \lesssim 1$ of the force calculation. The whole code is summarised by the flowchart of Fig. 2.12.

The approximation of the potential and the force by the tree structure influences the dynamical evolution of representative systems only slightly. Tests for open systems show that the energy conservation is only weakly violated, and the departure from exact conservation seems to be comparable to those errors typically tolerated from time integration errors. These errors are only weakly correlated from one timestep to the next, which leads to a fluctuation around a mean value rather than a steady growth or decay. Figure 2.13 shows an example of the energy conservation for a dense plasma system over 10,000 timesteps. The fluctuation in total energy is less than 0.5%.

Because most of the integrations are treated as particle–pseudoparticle interactions, the force calculation is not symmetric. This means that momentum and angular momentum are not exactly conserved. Empirical tests by Hernquist (1987) indicate that the departures are small for $\theta \lesssim 1$. In particular, the non-reciprocity in the force-law causes the centre of mass to drift, an effect which can be minimised by taking $\theta \leq 0.5$ and by increasing the number of particles.

Even in its most basic form, the hierarchical tree code is a very promising simulation tool and is already commonly used for the treatment of N-body problems. It can be implemented in a variety of programming languages like C, PASCAL, LISP, and FORTRAN. The advantage of C, PASCAL, and LISP is that they allow recursive function calls and permit a close correspondence between the tree structure and the program coding.

In a later chapter we will show how the computational performance of tree codes can be improved. On the software side, this means higher order integration schemes, vectorisation and introduction of individual timesteps. On the hardware side, there have been suggestions to design computers which are especially adapted to the tree structure. The tree algorithm can also be used on a parallel machine: Each particle has its own interaction list which can be summed independently.

3

Open Boundary Problems

In this chapter we will present examples of so-called open boundary problems. By this we mean that the simulation contains all particles relevant to the problem, and the size of the simulation region is adjusted accordingly at each timestep. This type of problem is by far the easiest one to which one can apply hierarchical data structures. The additional difficulties posed by periodic boundaries will be considered in Chapters 5 and 6.

3.1 Gravitational Problems in Astrophysics

Hierarchical tree codes were first developed in the context of astrophysics. This is not surprising because there is a big discrepancy between the number of bodies one would like to study – for example, $O(10^{11})$ for a galaxy – and the number one can afford to model with a standard N-body code – at present $O(10^5)$. PIC codes, which employ a grid structure to represent the fields in space, usually cannot handle these problems for two reasons: the complex structure of the investigated object and the large density contrasts such as those found in galaxies. N-body codes are able to avoid the first of these difficulties, because they are gridless and can therefore cope with arbitrarily complicated structures. The second difficulty remains, however, because of the N^2 scaling of computation time. This can have two consequences. If the number of simulation particles is too small, the spatial resolution of the simulation might not be good enough to reveal the real dynamic behaviour of the system. On the other hand, a system with a sufficient number of particles to give adequate resolution takes a long time to run, so one might not be able to follow the time-development for a long enough interval.

We saw in the previous chapter that hierarchical tree codes need less computation time for large N than particle–particle codes. Therefore, hierarchical tree codes permit us to use more particles in the simulation and run it for a

longer time. This increased resolution (in time or space) allows problems to be
investigated which were simply impossible to handle before.

Astrophysical applications range from star clusters, where the simulation
particle is equivalent to a single star, to large-scale structural features like clusters of galaxies, where simulation particles can represent a whole galaxy. In
general, the number of simulation particles N is not equivalent to the number
of constituents N_o, that is, stars, galaxies, etc., of the investigated object, but
one simulation particle represents N_o/N constituents. Of course, one wants to
have the ratio N_o/N as small as possible, but also big enough to include all important physical characteristics of the system. For example, if one is interested
in the core collapse that occurs during the dynamical evolution of a globular
cluster the following formation of binary stars and the ensuing expansion of
the core, a realistic N is thought to be $O(10^5)$ (Hut et al. 1988). Conventional
particle–particle codes can barely achieve that with the computational facili-
ties available at the moment, but tree codes have put such calculations within
reach. Having chosen an adequate ratio of N/N_o and the initial particle distribution, the dynamics of the system can then be described by the following set
of differential equations:

$$\frac{d^2\vec{x}_i}{dt^2} = \sum_{j,i \neq j} \frac{m_j(\vec{x}_j - \vec{x}_i)}{((\vec{x}_j - \vec{x}_i)^2 + \epsilon^2)^{3/2}}, \tag{3.1}$$

where the system of units is such that $G = 1$ and ϵ is a 'softening' parameter for
close encounters, which will be explained in what follows. In N-body simulations the solution of this differential equation system is replaced by a summing
of the forces for every particle. The softening parameter ϵ has to be introduced
for the following reason. If two or more particles are very close together, they
exert very strong forces on each other. These force terms are used to calculate
the new velocity and position in the next timestep t_{n+1}, which in their simplest
form are given by

$$v(t_{n+1}) = v(t_n) + F_n(t_n)/m * \Delta t, \tag{3.2}$$
$$x(t_{n+1}) = x(t_n) + v(t_n) * \Delta t. \tag{3.3}$$

The large force due to a close encounter leads to a strong change of the
velocities of the particles involved. In reality, this is balanced by the opposite
effect when the particles part again. If the timestep is too big, this can lead to an
imbalance in the description of the acceleration and deceleration and big errors
in the new positions and velocities of the particles involved. Fortunately, in many
cases, close encounters between particles are so rare that the exact description
of close collisions can be neglected. In such 'collisionless' situations one avoids

big force terms by introducing a large softening parameter ϵ in Eq. 3.1. In most collisionless simulations this parameter ϵ is chosen to be of the order of the average interparticle spacing.

For collisional problems it is obviously not desirable to modify the force in such a way. The timestep has to be small enough to avoid large errors of the positions and velocities during close encounters. Choosing a small timestep makes the simulation more expensive: In Chapter 4 we consider some ways to get around this problem. However, it is necessary to choose an appropriate softening to avoid a breakdown of the numerics, and unphysical particle orbits. To achieve that, one can use either the force law as Eq. 3.1 and a very small ϵ or an altogether different force law.

How does one decide whether a collisional or collisionless treatment is appropriate for astrophysical applications? There are two characteristic timescales in these types of N-body problems – one is the average time t_{cr} a particle takes to cross the characteristic size of the system, the other is the two-body relaxation time t_{rel}, which is the average time for a particle to change its energy by successive collisions. The ratio of these two timescales is approximately (Spitzer Jr. 1987)

$$\frac{t_{rel}}{t_{cr}} \sim \frac{N}{11 \ln(0.4N)}. \tag{3.4}$$

Comparing the intended simulation time t_{sim} with these two characteristic times of the system allows us to determine whether collisions play an important role in the physical behaviour and if it has to be included in the simulation or not. If $t_{sim} \sim t_{cr}$, the system can be regarded as collisionless, whereas if $t_{sim} \sim t_{rel}$ it is collisional. This means a system is not intrinsically collisional or collisionless: It all depends on the timescale of observation. For example, the evolution of a binary globular cluster observed in Magellanic clouds into a single globular cluster takes about 10^7 years and can be simulated as a collisionless system (Sugimoto & Makino 1989). However, the cluster contracts gravothermally and binary star systems are formed over a timescale of 10^9 years (Elson et al. 1987). For this process, two-body relaxation becomes important and the simulation must include collisions. This means the same object needs collisionless or collisional treatment depending on the phase of its evolution.

In order to represent a collisionless system faithfully, the simulation method must minimise relaxation effects. Due to particle discreteness, two-body relaxation shows up as irregular fluctuations in the potential energy of the system. Hernquist and Barnes (1990) showed that such fluctuations occur in tree-code simulations of collisionless astrophysical systems, which are comparable to the fluctuations from Monte Carlo sampling and scale like $N^{1/2}$. There has been some discussion whether relaxation would make tree codes unsuitable for the simulation of collisionless systems and that other codes are more suitable for such systems (van Albada 1986, Sellwood 1987). However, Hernquist and Barnes (1990) showed that relaxation proceeds at nearly the same rate as in particle–particle codes and that contrary to the prevailing view, expansion techniques and particle-based codes have fluctuations of the same order. Therefore, tree codes are as good as any other method for simulating collisionless systems and indeed have mainly been used in this context.

Having considered how to simulate a given astrophysical problem, we now discuss some examples that have been investigated using tree codes. The majority of simulations have been carried out in the context of the dynamics of galaxies. This is not surprising because tree codes were first developed by people working in this field and there are many open questions concerning the dynamics. What makes N-body simulations – and especially tree codes – so attractive for the study of galaxies is that no special difficulties are posed by the variety of sizes and shapes and the strong density contrasts present in these objects.

Most galaxies have some ordered structure and can be classified as either elliptical or spiral. Elliptical galaxies are oval swarms of stars distributed in complicated three-dimensional orbits, whereas spiral galaxies are flattened disks in which all stars orbit in the same direction about a common centre. However, some galaxies do not belong in either of these categories, but have highly irregular features. The interpretation of such seemingly unstructured galaxies has been a matter of controversy for a long time. Computer models combined with increasingly powerful observations (Schwarz 1986) have provided new insight into this debate, and there are now good reasons to believe that many peculiar systems could be ordinary galaxies undergoing collisions. One such system – 'The Antennae' (NGC 4038/39) – exhibits many similarities with a simulation involving two disk galaxies (see Fig. 3.1).

Simulations with test particle codes were first performed by Toomre and Toomre (1972). However, at that time the limitations in computation time prevented examination of the long-term evolution of galaxy collisions and it could not be proved that such collisions lead to mergers and the formation of galaxies of a different type. How are tree codes now used to model collisions of galaxies?

Most of the visible matter of typical spherical galaxies exists in the form of stars and only about 10% as atomic hydrogen gas. For elliptical galaxies, the proportion of atomic hydrogen is even less, so that to lowest order it can be neglected in a simulation. (On the other hand, elliptical galaxies may contain large quantities of ionized gas.) Taking the Milky Way as an example, it can be shown that the gravitational force exerted on the sun by the nearest known

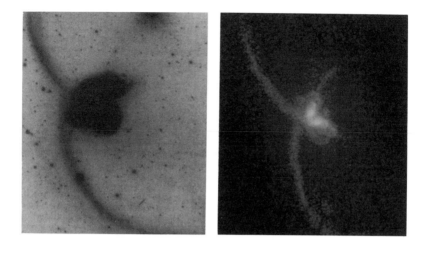

(a)   (b)

Fig. 3.1. Comparison between observed galaxy and simulation. (a) Optical photo-graph showing 'The Antennae' a nearby pair of interacting galaxies, courtesy of F. Schweizer, Carnegie Institution of Washington. (b) A self-consistent model resem-bling the Antennae. The simulation started with a pair of disk galaxies placed on an elliptical orbit; after their first passage, extended tails developed from both disks and the view is edge-on to the orbital plane. The grey stars represent dark matter. Simula-tion graphic reproduced by courtesy of J. Barnes from Barnes and Hernquist (1993). ©1993 American Institute of Physics.

star is 10^3 times smaller than the collective force exerted by the rest of the galaxy. Therefore, the system can be assumed to be collisionless. Dynamical studies indicate that in addition to the stars and gases of the disk and the central bulge, a great deal of invisible matter is present in spiral galaxies which might be as much as five times the mass of the visible matter. It is still not known what this invisible matter actually is. It might consist of burned out stars, black holes weighing up to 10^6 solar masses, or even subatomic particles (Tremaine 1992). For simulation purposes this means that the simulation particles have to represent not only stars *but dark matter as well*, which, like the stars, can be modelled as collisionless bodies (Barnes & Hernquist 1993).

By contrast, if one wants to include the gaseous constituents of the galaxy, one must include the pressure and viscous forces that the molecules and atoms generate. These must be treated separately by a hydrodynamic code (Benz 1988), although there also exist hydrodynamic codes which make use of the tree structure (Hernquist 1988, Hernquist & Katz 1989). To perform numerical experiments of the formation of galaxies, the initial particle positions can be set up using the cosmological spectrum evolved according to linear theory, as described by Efstathiou et al. (1985), and Dubinski and Carlberg (1991). Setting up the particle distribution of already-formed galaxies, including the dark matter halos, luminous disks, and bulges, and the interstellar gas is much more complicated. For a discussion of the various procedures and their effects on the simulation results see Hernquist (1993a,b).

To study the dynamics of colliding galaxies, these simulations focus on problems which involve large-scale distributions where the detailed structure is probably not critical, but one has to be aware that the detailed behaviour of a galaxy is a good deal more complicated than the ideal gas model adopted in N-body particle simulation. Other more complex effects, such as magnetic fields, cosmic rays, star light, interconversion between gas and stars, and the multiphase nature of the galaxies' constituents, are usually neglected.

The outcome of the encounter of two galaxies depends very much on the relative velocity of the galaxies involved. If the relative velocity is high the two galaxies just pass each other without disturbing each other very much, but for lower velocities the situation is quite different: Barnes and Hernquist (1993) showed the simulation of a collision between two identical spiral galaxies similar in mass and structure to the Milky Way (Fig. 3.2). Both galaxies were modelled as 44,000 mass points, some representing the dark matter, others the stars. Each model galaxy contained a compact central bulge, a thin spinning disk and a spheroidal halo containing 80% of the total mass. The two galaxies were placed on approaching orbits with speeds that would be acquired by falling towards each other from an infinite distance. The orbits were chosen so that the encounter was not quite head on, and the disks were tilted against each other. After the collision the galaxies did not separate to an infinite distance after their passage, but after having reached a maximum separation they fell back together for a much closer collision and merger.

The explanation for this behaviour is that particles orbiting within each halo in the same direction acquire energy and angular momentum at the halo's relative motion. Consequently, the relative orbits decay and the halos fall back together for ever closer encounters, then the luminous components of each galaxy lose orbital momentum with its own halo. The merger remnants have smooth and nearly featureless stellar distributions with three-dimensional surfaces of constant stellar density approximating triaxial ellipsoids (Barnes & Hernquist 1993, Hernquist 1993b). These properties are consistent with those of elliptical galaxies. This implies that elliptical galaxies could be formed through encounters of spherical galaxies.

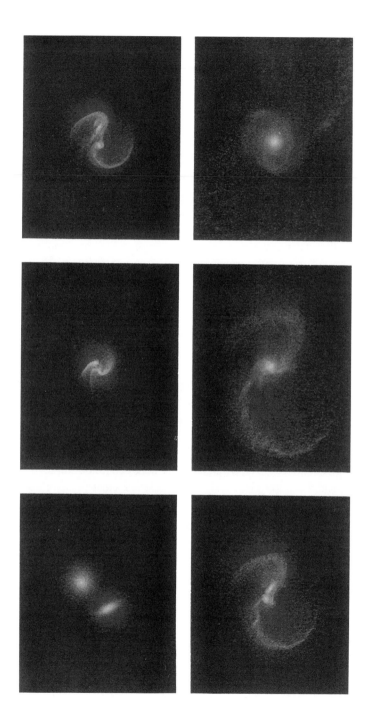

Fig. 3.2. Simulation of the dynamics of the collision and merger of two identical galaxies. Reproduced by courtesy of J. Barnes from Barnes and Hernquist (1993). ©1993 American Institute of Physics.

If the two interacting galaxies are of unequal size, the gravitational field of the more massive galaxies tends to bend and fold the form of the flattened disk of the smaller galaxy into a curved surface. Each time the smaller galaxy wraps around the bigger one, it produces another thin curved region of stars. Recent studies of such 'satellite mergers' can be found, for example, in Hernquist and Mihos (1995). Figure 3.3 shows a typical simulation in which a satellite galaxy was placed on an orbit inclined 30° to a larger disk galaxy. In the first five frames the satellite orbit decays rapidly owing to friction with the dark halo surrounding the disk. Tidal forces from the satellite then excite spiral arms and a central bar mode in the larger galaxy, which eventually dissipate via mixing. Seen from Earth the edges of the surface appear relatively thick and bright.

However, there are some features that cannot be explained this way. For example, some elliptical galaxies have counter-rotating stellar cores and bright elliptical galaxies rotate slowly, whereas faint ones do not. Including the gas dynamics in such calculations could perhaps explain these phenomena. For further reading on galaxy interaction, see Hernquist and Quinn (1988), Barnes, Hernquist and Schweizer (1991), and Wielen (1990).

Another open question is that of 'active' galaxies. In the 1980s the Infrared Astronomy Satellite found galaxies with exceptional infrared luminosity, which is suggestive of star formation. Young stars form within gas clouds, which absorb their visible light and reemit it in the longer wavelengths. Modelling the formation and evolution of individual stars within the simulation is out of the question, but phenomenological approaches that convert gas into stars using probability rules derived from star forming regions in our own galaxy may prove effective.

McMillan and Aarseth (1993) showed how tree codes could be employed to investigate *collisional* astrophysical systems: namely star clusters and galactic nuclei. In such collisional systems a high accuracy over many system crossing times is required, because close encounters and small-scale bound subsystems – so-called binaries – can have a considerable effect on the entire system.

For the simulation of galaxies just discussed, an accuracy of about 1% of the total energy over a few crossing times is considered sufficient. By contrast, for star clusters accuracies of $|\Delta E/E| \ll 0.04 N^{-1} t/t_{cr}$ are needed over several tens of *relaxation times*, which for a system of 10^3 stars would mean an accuracy of 10^{-5}. In order to achieve such a high accuracy McMillan and Aarseth (1993) chose $\theta \leq 0.5$ and a multipole expansion including octupole terms.

The choice of the timestep needs special care in such collisional systems. The reason is that although only 100 timesteps per crossing time are usually sufficient, the few stars which are involved in bound subsystems need a much smaller timestep for a correct evaluation of their dynamics. The orbital period

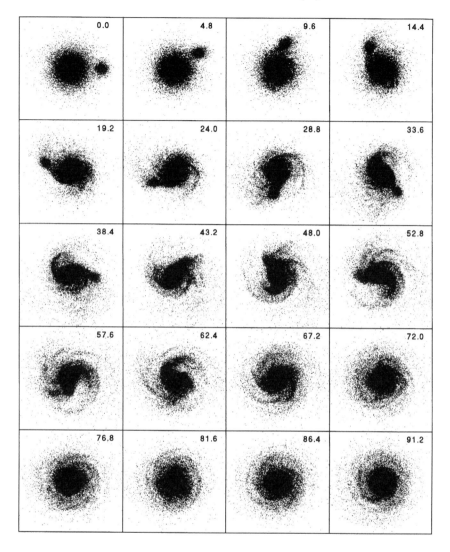

Fig. 3.3. Evolution of a stellar disk in a satellite merger. The large galaxy comprises 90% luminous material and 10% gas and is surrounded by a dark halo roughly six times the total galaxy mass. The mass ratio between the disk and satellite galaxies is 10:1. Reproduced by courtesy of Hernquist and Mihos (1995). ©1995 the American Astronomical Society.

of binaries is given by $t_{bin} = 2^{1/2}\pi t_{cr}/N$ and therefore the required timestep is a factor $\sim N^{2/3}$ shorter than that which would be necessary for the rest of the system. Forcing all particles to have the same small timestep would be very inefficient, therefore a block timestep algorithm was used in this simulation (McMillan & Aarseth 1993). A more detailed discussion on how to improve the accuracy by using different timesteps and high-order integration schemes can be found in Chapter 4.

Tree codes have also been used in astrophysics with periodic boundary conditions, which will be discussed in different contexts in Chapters 5 and 6 of this book. For example, the observation that the universe possesses large density contrast on scales smaller than the correlation length has been investigated by Suginohara et al. (1991) and Bouchet and Hernquist (1992). It is widely believed that this structure arose through gravitational clustering from relatively smooth initial conditions as the universe expanded following the big bang, and different models are tested by means of tree code simulations.

Periodic tree codes have also been used to investigate gravitational problems on a much smaller scale – planetesimal dynamics. Planetesimals are investigated in the context of the formation of terrestrial planets. Current theories assume four different processes at work: (a) the condensation of heavy elements into volatile and refractory grains (Nakagawa et al. 1986), (b) the formation of planetesimals (Goldreich & Ward 1973), (c) the coagulation of planetesimals into protoplanets (Wetherill & Stewart 1989, Aarseth et al. 1993), and (d) clearing of the residual planetesimals. There exist many analytical approaches on the evolution of planetesimals (Safronov 1969, Greenberg et al. 1978, Wetherill & Stewart 1989), but as Aarseth, Lin, and Palmer (1993) pointed out, in order to address the most central issues in planetesimal dynamics and planetary formation, numerical simulations of systems with large dynamical ranges and high spatial resolution are required to include effects such as gravitational interactions between particles, physical collisions, and particle spin. Richardson (1993) used a tree code with specially adjusted periodic boundary conditions to investigate the dynamics of planetesimals.

In summary, it can safely be said that within the last ten years, tree codes have developed from a novel method for reducing the computation time of N-body calculus, to becoming one of the standard techniques for studying galactic dynamics. This success owes much to the fact that explanations to many open questions in this context were found by performimg such simulations with tree codes. On the other hand, applications of tree codes to collisional and periodic problems are still relatively new, as are applications in other fields outside the field of astrophysics. Two such examples will be discussed in the following sections of this chapter.

3.2 Space-Charge Dominated Particle Beams

The subject of charged particle beam dynamics is too vast to do justice to within the few pages we devote here to this topic, and we will not attempt to describe all the issues and design considerations which go into modern linear and circular accelerators, nor the numerous applications of particle beams. Nonetheless, it is helpful to review a few basic concepts in order to identify the types of beam propagation problems which might usefully be tackled by tree codes. More complete introductions to particle beam physics can be found in Lawson (1988) and Wiedemann (1993).

Beams come in many shapes and sizes, depending on how they are created – from electron or ion sources – how they are accelerated and transported, through which medium they are propagated, and onto what object they are eventually targeted. Despite this variety, there are a number of general definitions and quantities common to most types of beam. The most important parameters are the beam energy or momentum, the current, and emittance, or the particle density in (6-dimensional) phase space.

The beam energy is simply the kinetic energy of a single particle in the beam, which also carries a charge $-e$ (electron) or Ze (ion). For ions, this energy is usually defined as the kinetic energy per nucleon. The current is essentially the charge passing a given point per unit time. For pulsed beams consisting of a regular series of particle bunches, this definition depends on the time interval used, so one normally distinguishes between the peak current: $I_p = Q/\tau$, and the average current: $I_{av} = Q/\tau_r$, where Q is the total charge of a single particle bunch, τ is the bunch length, and τ_r is the time interval between pulses. The current is often expressed in terms of the beam intensity or particle flux Φ:

$$I = eZ\Phi.$$

In circular accelerators, this becomes:

$$I = eZfN,$$

where f is the revolution frequency and N is the total number of beam particles.

One of the most useful quantities to characterise beam behaviour is the emittance. A beam can generally be described completely by the time-evolution of each particle in 6-dimensional phase space (x, y, z, p_x, p_y, p_z). According to Liouville's theorem, the density of particles in phase space stays constant under the influence of conservative external fields – see, for example, Wiedemann (1993). Furthermore, provided there is no coupling between the three spatial degrees of freedom, each *pair* (x, p_x), (y, p_y), (z, p_z) is individually conserved.

A finite-temperature beam can thus be described by 3 independent phase-space planes, the areas of which are conserved along the trajectory. The emittance in one of these planes – say (x, p_x) – is defined as:

$$\pi\varepsilon = \int \frac{dx\,dp_x}{p_0}, \tag{3.5}$$

where $p_0 = mc\sqrt{\gamma^2 - 1}$ is the average particle momentum, and generally $p_0 \gg p_x, p_y$. The transverse emittance is usually expressed in units of mm rad or π mm rad according to context and convention.

Although the emittance is a constant of motion, the shape of the phase-space region occupied by the particle will change during propagation, for example, as a result of focussing or defocussing in an accelerator. One can develop this further mathematically by formally enclosing the particles within an ellipse whose evolution depends on the external focussing elements in the beam line. Particles sitting on the edge of this ellipse at, say, $x = y = a$ obey a beam envelope equation which depends only on ε and a function $K_o(z)$ representing the focussing and acceleration elements. In the absence of acceleration – a situation which might exist, for example, in a storage ring – we can set $d\gamma/dz = 0$, and the beam evolves according to:

$$\frac{d^2 a}{dz^2} + K_o a - \frac{\varepsilon^2}{a^3} = 0. \tag{3.6}$$

An important assumption made in using Eq. 3.6 is that the three components of phase space decouple. This is true provided that the magnetic focussing fields are azimuthal (or transverse to the plane or propagation), and that the electrostatic forces between the particles themselves are negligible. The first requirement is easily met by typical focussing elements, for example, quadrupole magnets, but the second condition is often not satisfied for ion beams whose particles carry a high charge Z. Transport of such beams poses additional problems because the space-charge effects can cause the beam to blow apart, or to go unstable.

One can determine the magnitude of the self-fields with simple physical arguments as follows (Lawson 1988, Wiedemann 1993). Suppose that the beam has a uniform particle distribution ρ and a radius a. It will exhibit a self-electric field E_r and a magnetic field B_ϕ due to the charge and current carried by the particles. Applying Coulomb's law $\nabla \cdot E = 4\pi\rho$ in cylindrical coordinates we obtain:

$$E_r = 2\pi\rho r. \tag{3.7}$$

Likewise, from Ampere's law $\nabla \wedge B = \frac{4\pi}{c}\rho v$, it follows that:

$$B_\phi = 2\pi\rho \frac{v}{c} r. \tag{3.8}$$

A particle within the beam will thus feel an outward electrostatic force due to E_r and an inward 'pinch' force due to B_ϕ. The net Lorentz force is therefore:

$$F_r = e\left(E_r - \frac{v}{c}B_\phi\right)$$

$$= 2\pi e\rho r\left(1 - \frac{v^2}{c^2}\right)$$

$$= 2\pi e\frac{\rho}{\gamma^2}r. \tag{3.9}$$

This result says that the repulsive electrostatic field is compensated by the magnetic field to $O(1/\gamma^2)$, that is, the net force vanishes at high energies. Thus, for 'bare' relativistic electron beams this effect can generally be neglected, though one should note that the cancellation is incomplete for beams in conducting pipes. For slow (nonrelativistic) ion beams, however, the charge density is a factor Z higher and $\gamma \rightarrow 1$. Under these conditions, the beam is said to be *space-charge dominated*, and additional focussing measures have to be applied to maintain stable beam propagation over long distances.

Letting $\rho = ZeN/\pi a^2$, where N is the line number density, Eq. 3.9 leads to an envelope equation for a beam expanding under its own space-charge:

$$a\frac{d^2a}{dz^2} = \frac{2N\mu r_c}{\beta^2\gamma^3} = K_s, \tag{3.10}$$

where $r_c = e^2/m_ec^2$ is the classical electron radius, $\mu = m_e/M_i$ is the ratio of electron mass to ion mass, and $\beta = v_z/c$ is the normalised longitudinal beam velocity. Equations 3.6 and 3.10 can be combined to give the envelope equation of an axisymmetric beam including external forces and space-charge:

$$\frac{d^2a}{dz^2} + K_oa - \frac{K_s}{a} - \frac{\varepsilon^2}{a^3} = 0. \tag{3.11}$$

This equation can be generalised to the more practical 2D case (i.e., where the beam is nonaxisymmetric) to yield the well-known Kapchinskii–Vladimirskii (K–V) equations (Kapchinskii & Vladimirskii 1959):

$$\frac{d^2a}{dz^2} + K_xa - \frac{2K_s}{a+b} - \frac{\varepsilon_x^2}{a^3} = 0$$

$$\frac{d^2b}{dz^2} + K_yb - \frac{2K_s}{a+b} - \frac{\varepsilon_y^2}{b^3} = 0, \tag{3.12}$$

where a and b are now the beam·radii in the x and y directions, respectively; ε_x and ε_y are the corresponding emittances. More complete versions of the K–V equations can be found in Lee and Cooper (1976) and Borchardt et al. (1988).

A beam with a K–V distribution in phase space can in principle be matched to an arbitrary set of focussing and accelerating elements, but in practice it tends to be unstable. Resonances between the beam and a periodic focussing lattice can lead to equipartition between the x and y planes, emittance growth, and the formation of haloes. Another complication arises for bunched beams, which excite longitudinal wakefields, which in turn result in dynamic coupling between adjacent bunches (Chao 1983). In many applications there is therefore a large uncertainty in predicting the final brightness of the beam. Much effort has recently gone into 3-dimensional particle-in-cell simulation to address transport and stability issues of space-charge dominated beams. In particular, extensive studies have been made by groups at Livermore (Friedman et al. 1992), Berkeley (Fawley et al. 1993), and NRL (Haber et al. 1991) to evaluate the feasibility of using heavy ion beams as a potential driver for inertial confinement fusion (ICF).

One of the more sophisticated codes developed to date is the WARP code (Friedman et al. 1992), which includes special features to describe bends, self-field boundary conditions in conducting pipes, and discrete accelerator elements with sharp boundaries. This code has been successfully used to tackle some of the major design issues – most of which boil down to improving beam quality, that is, achieving high current whilst minimising emittance growth. A summary of these studies can be found in Friedman (1992, 1993).

Given the impressive achievements already made by PIC simulation, what can we hope to gain by using a tree code for beam modelling? A problem common to all PIC codes is the presence of numerical heating, which arises due to the way in which particles interact with the grid (aliasing) (Birdsall & Langdon 1985). In heavy ion simulation, the instability can occur particularly for beams which are initialised colder in the longitudinal direction than in the transverse, that is, $T_z \ll T_{x,y}$. In this case, 'equilibration' takes place, which heats the beam in z until the temperatures are of the same order. Although this is thought to be a physical effect, numerical heating will generally cause some temperature increase (and hence emittance growth) unless the Debye length is spatially resolved. This makes it difficult to separate the physical effects from the nonphysical without increasing the computational costs (Friedman et al. 1992).

This problem could be tackled efficiently with a tree code, which thanks to its gridlessness, does not suffer from numerical heating – at least not to the same extent; see, for example, Ambrosiano et al. (1988). Paradoxically, the better the beam quality, the harder it is to model with PIC because the temperature is smaller, and the Debye length is harder to resolve. A tree code also has the advantage that accelerator elements and bends can be introduced without substantial modifications to the algorithm or the special warped meshes required in a PIC code.

An issue which we have not mentioned so far is the presence of Coulomb scattering within the beam. In deriving the envelope equations (3.12), it was assumed that the particles were essentially collisionless, and did not suffer binary interactions. The space-charge effect leads to additional defocussing which can in principle be corrected by adding lenses: It does not necessarily limit the maximum obtainable brightness of a tightly focussed beam. Binary scattering, on the other hand, is a non-Liouvillian effect which causes longitudinal energy broadening – the so-called Boersch effect (Boersch 1954) – and lateral position and velocity shifts, which can both lead to blurring of the final focus. These effects are especially important for very intense beams used in electron lithography, scanning microscopy, and ion implantation, for which both high current and high resolution are often required. Modelling of collisional beams has been tackled using both approximate analytical theory and Monte Carlo simulation (Jansen 1990), where the latter numerical approach is limited to relatively low numbers of particles. The speedup gained using a tree code would provide a means to test focussing systems at the design stage. The ability to simulate beams with a large number of particles would allow collisional and space-charge effects to be investigated simultaneously.

The holy grail of phase-space density maximisation or beam cooling is to produce and maintain a *crystalline* beam, for which the focussing forces exactly compensate the space-charge, thus avoiding emittance growth altogether (Rahman & Schiffer 1986). Particle-in-cell simulation is inadequate to model crystalline beams because they are strongly coupled. Their temperature is reduced to such an extent that their potential energy (due to space-charge) vastly exceeds their thermal energy. This problem is closely related to that of strongly coupled plasmas, which we discuss in Chapter 6. Molecular dynamics studies have been previously performed for somewhat idealised systems (Rahman & Schiffer 1986, Hasse 1991), but as usual these were limited to a few thousand particles. Realistic beams, on the other hand, contain around 10^9 particles. More recent studies have considered the requirements for maintaining crystalline beams in realistic storage rings (Li & Sessler 1994).

3.3 Collisions of Heavy Nuclei: A Case Study

The study of nuclear matter is challenging, both theoretically and experimentally, because of the high energies needed to gain information about its properties. Although not immediately accessible to everyday experience, nuclei (neutrons and protons) make up the core of atoms, and are the main constituent of neutron stars. Just like for other materials – fluids, solids, plasmas, etc. – the properties of nuclear matter can be described by a nuclear equation of state. This is of fundamental importance for the understanding of phenomena such as the stability of neutron stars, supernova collapse, and the formation of the early universe.

The main experimental tool presently available to study nuclear matter is heavy ion collisions. At sufficiently high energies (typically tens of MeV per nucleon), fully stripped ions can approach each other to within a few Fermi (1 fm $= 10^{-15}$ m), causing the nuclei to interact and break into fragments. The energy, mass, and angular distribution of these reaction products yield information about the equation of state and transport coefficients. A comprehensive introduction to this subject can be found in the recent book by Csernai (1994).

Generally, one distinguishes between three types of heavy ion interaction according to the beam energy: (i) intermediate nuclear reactions (10–100 MeV/ nucleon), (ii) relativistic reactions (100 MeV/nucleon – 10 GeV/nucleon), and (iii) ultrarelativistic reactions (> 10 GeV/nucleon). These energies correspond to different regimes for the equation of state. In the first of these regimes, nuclear matter remains close to its usual density and exhibits multifragmentation, forming 'droplets' of nuclear liquid during an interaction. The second regime allows matter to be studied under similar conditions to those found in neutron stars and supernovae. Properties of the nuclear equation of state derived from heavy ion models and experiments can be compared directly with observations of astrophysical objects. In the third regime, it is thought that 'quark-gluon plasmas' will form – an exotic state of matter in which quarks and gluons exist unbounded for a short period of time (< 10 fm/c).

The first models of heavy ion collisions were based on nuclear fluid dynamics (NFD) (Csernai 1994), which can describe collective flow, and inelastic collisional effects such as the production of pions (π^{+-}, π^{0}) during the interaction. This approach, however, cannot treat fragment formation or nonequilibrium states. A more advanced microscopic theory is the Vlasov–Uehling–Uhlenbeck (VUU) model (Cassing & Mosel 1990, Lang et al. 1992). This approach, which is closely related to Vlasov–Fokker–Planck modelling of plasmas (see Chapter 6), replaces the discrete nucleons by a smooth 6-dimensional distribution function $f(\mathbf{r}, \mathbf{p})$. A major advantage of the VUU model is that not only can nonequilibrium (or 'kinetic') effects be followed, but quantum effects are included explicitly. In particular, the exclusion principle is obeyed through a 'Pauli blocking' term, which ensures that no overlapping occurs in phase space. Despite the successful application of this model to describe a number of phenomena, it does not include the many-body correlations needed to correctly model fragmentation.

The main alternative to VUU is the so-called Quantum Molecular Dynamics technique (QMD) (Aichelin & Stöcker 1986, Neise et al. 1990). Just as in classical molecular dynamics, nucleon trajectories are followed explicitly by integrating their equations of motion. Nucleons are represented by 6D-Gaussian wave packets, which interact via a number of potentials representing different physical effects.

The main disadvantage of QMD is its computational expense. Nucleon trajectories are integrated by determining the force on each particle due to all the others, which, like most N-body problems, incurs a cost of $N(N - 1)/2$ computations, and hence a CPU-time scaling of N^2 for large N.

In this section, we will show how to apply the tree method to the QMD model in order to reduce this scaling to $N \log N$, thereby yielding considerable savings in computing time for heavy nuclei systems where $N \geq 200$ (Gibbon 1992a). The price to be paid is, as usual, a small controllable error in macroscopic quantities such as the total energy. Before we describe in detail how it can be adapted for QMD, it is helpful to briefly recall the tree algorithm to see how it fits into the context of heavy nuclei collisions, and to introduce some of the nuclear physics terminology.

As we saw in Chapter 2, the underlying philosophy of the tree method is to reduce the number of terms required in the force summation for each particle. This is done by adding contributions from the nearest neighbours, but grouping distant particles into larger 'pseudoparticles' and including their contribution as a single, collective term, centred at the group's 'centre of mass'. The term mass is used here somewhat loosely, and has physical significance only where gravitational forces govern the interaction. The equivalent in the QMD context depends on the precise nature of the force, and we will see shortly that there are several different quantities (or charges) in the model which determine the size and 'sign' of the pseudoparticles.

First, a tree structure is built around the nucleons. This is done by surrounding the nuclei (usually 2 – target and projectile) and then recursively subdividing this volume until there remains only one nucleon (or 'leaf') per cell.

Next, the cells are regrouped from the leaves down to the root to form successively larger pseudoparticles. The interaction list for each particle is obtained by applying the usual acceptance or tolerance criterion, starting at the root and

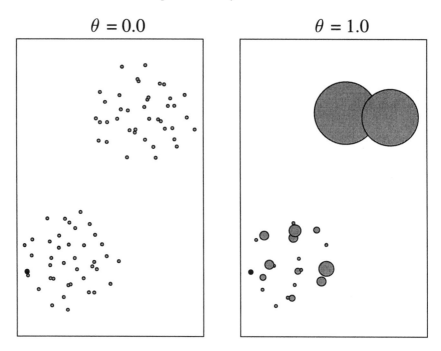

Fig. 3.4. Interaction lists for a target nucleon (X) with (a) $\theta = 1.0$, (b) $\theta = 0.0$. The radii of the pseudoparticles are proportional to the number of nucleons they contain.

working up to the leaves: If $s/d < \theta$, then add the term to the list and move to the next node (sister, uncle, etc.); otherwise, resolve the cell and repeat the test on its first 'daughter' node. The tolerance parameter θ is typically 0.3–1.0 depending on the desired accuracy (or speed).

The result of this procedure is illustrated in Fig. 3.4, which shows the interaction list for one of the target nucleons. A physically appealing feature with this choice of θ is that initially this nucleon sees no detail within the projectile; it interacts with the nucleus as a whole. One should note, however, that the groupings are derived purely from the geometrical procedure described previously, and bear no relation to the physical nuclei fragments which may form during the course of the interaction.

Comparing the $\theta = 1.0$ list to the one for $\theta = 0$ (in which we recover the direct particle–particle sum), we see that the number of terms has been reduced from 79 to 20, with obvious potential for timesaving in the force calculation. As we shall see later, however, a number of other factors conspire to degrade the performance of the tree algorithm in the QMD application relative to the

PP direct sum, but we still achieve significant savings for the heavier systems, in which the total number of protons and neutrons, $N \sim 400$.

There are a number of physical features unique to QMD which deserve some additional consideration when designing a tree code for this application. In the version of the Frankfurt code on which the algorithm here is based (Konopka 1992), there are a total of five different interactions which occur between the nucleons: Coulomb, Pauli, Skyrme 2-body, Skyrme 3-body, and Yukawa. Nucleons are represented by overlapping Gaussian wave packets, whose centres propagate according to classical kinematics. In the 'fixed-width' mode of this model, the packets remain rigid: an assumption which allows the 3-body interaction sum over $O(N^2)$ terms to be reduced to a sum over $O(N)$ – as for the other four interactions. Each potential therefore has an $O(N^2)$ CPU-time scaling. We do not consider the 'variable-width' mode here, but note in passing that the N^3 CPU scaling of the Skyrme 3-body term could in principle be reduced to $N \log^2 N$, provided that appropriate pseudoparticle pairs can be constructed for this interaction.

In the fixed-width mode, the five potentials can be classified into three groups:

- Yukawa, Skyrme 2-body, and Skyrme 3-body, which permit all possible interactions between neutrons and protons (n-n, n-p, p-n), regardless of spin or isospin
- Coulomb, which allows interactions between protons only (p-p)
- Pauli, which permits interactions only between particles with the same isospin ($\tau = 0, 1$)

The potentials are given by (Neise et al. 1990):

$$V_{ik}^{Yukawa} = \frac{1}{2r_{ik}} K_Y \left[\exp(-r_{ik}/\gamma_Y)\mathrm{erfc}A^- - \exp(r_{ik}/\gamma_Y)\mathrm{erfc}A^+ \right]$$

where

$$A^\pm = \frac{1}{2\gamma\gamma_Y\alpha^{\frac{1}{2}}} \pm \alpha^{\frac{1}{2}}r_{ik},$$

and γ_Y is a constant and $L = (2\alpha)^{-1/2}$ is the packet width.

$$V_{ik}^{Skyrme-2} = K_{S2}\exp(-\alpha r_{ik}^2)$$

$$U_i^{Skyrme-3} = K_{S3}\left(\sum_k \exp(-\alpha r_{ik}^2)\right)^\gamma \quad \text{(energy)}$$

$$V_{ik}^{Coulomb} = K_C \tau_i \tau_k \text{erf}\left(\alpha^{\frac{1}{2}} r_{ik}\right) / r_{ik}$$

$$V_{ik}^{Pauli} = K_P \tau_{ik}^{pau} \exp\left(-\frac{p_{ik}^2}{2p_o^2} - \frac{r_{ik}^2}{2q_o^2 + \frac{1}{\alpha}}\right),$$

where

$$\tau_{ik}^{Pauli} = (1 - |\tau_i - \tau_k|)\frac{1}{2} |\sigma_i + \sigma_k|,$$

K_Y, K_{S2}, K_{S3}, K_C, and K_P are numerical constants, and σ_i, σ_k are the particle spins ($= \pm 1$).

The first three potentials have no spin (or 'charge') factor, which means that pseudoparticles can be constructed just by summing all the nucleons within a node of the tree, and centering it with equal weighting. Referring to the schematic representation of the multipole expansion in Fig. 2.10, the potential of a pseudoparticle at $\mathbf{R}$ as seen by particle p at the origin is given by:

$$V(\mathbf{R}) = \sum_{i=1}^{m} V(\mathbf{R} - \mathbf{r}_i)$$

$$= \sum_i \left(1 - \mathbf{r}_i \frac{\partial}{\partial \mathbf{r}} + \frac{r_i^2}{2} \frac{\partial^2}{\partial r^2} \cdots\right) V_0 \exp(-\alpha R^2)$$

$$\simeq m V_0 \exp(-\alpha R^2)$$

$$+ 2\alpha V_0 R \exp(-\alpha R^2) \sum_s r_s$$

$$+ \ldots .$$

The dipole term vanishes if we take $\mathbf{R} = $ 'centre of mass' of the particle group, since then $\sum_i r_i = 0$. The short range of the three dominant potentials (Skyrme and Yukawa) allows us to truncate the expansion at the monopole term. Quadrupole terms could be included in principle, but would add considerably to the tree-code overhead in comparison to a direct particle–particle sum, and does not seem worthwhile for an application in which $N < 500$. Physically, we make the approximation that a group of wave packets may be replaced by a single packet of width α, and an amplitude equal to the number of nucleons in the cluster (m in the previous example). The same applies to the Skyrme 3-body and Yukawa interactions.

The Coulomb potential can be treated similarly, except that we form pseudoparticles of protons only. The centre in this case is given by:

$$\mathbf{r}_c = \frac{\sum_i \tau_i \mathbf{r}_i}{\sum_i \tau_i}.$$

The Pauli potential presents two additional complications. First, the spin factor τ_{ik}^{pau} is a function of the nucleon for which the force is being evaluated, as well as the nucleons within the pseudoparticle. The interaction occurs only between nucleons of like isospin τ, but there are four combinations of τ and σ which can satisfy this. For example, a neutron with $\tau_i = 0, \sigma_i = \pm 1$ can interact with another with $\tau_k = 0, \sigma_k = \pm 1$; a proton ($\tau_i = 1, \sigma_i = \pm 1$) can interact with another proton ($\tau_i = 1, \sigma_i = \pm 1$). This means that four sets of Pauli pseudoparticles are needed to cater to each of these possibilities. The spin factors of the pseudoparticles are given by:

$$\tau_{\pm 1}^{Pauli,0} = \sum_{neutrons} \frac{1}{2} |\sigma_s \pm 1|$$

$$\tau_{\pm 1}^{Pauli,1} = \sum_{protons} \frac{1}{2} |\sigma_s \pm 1|.$$

The pseudoparticle finally chosen will depend on the particle it interacts with, that is, (τ_i, σ_i).

The second problem with the Pauli interaction is the momentum dependence. The relative momenta of two particles is just as important as their spatial separation in determining the strength of the force between them. We therefore require a *6-dimensional* tree to obtain a suitable interaction list (i.e., pseudoparticles must be grouped in momentum-space as well as configuration-space). In principle, this presents no greater difficulty than extending the tree structure from, say, 2D to 3D, but does demand some extra bookkeeping, and a modified acceptance criterion for the interaction list. Instead of $s/d < \theta$, we use:

$$\frac{sq}{\Delta r \Delta p} < \theta_p^2, \tag{3.13}$$

where s, q are the space and momentum cell sizes, Δr and Δp are the separations between particles in space and momenta, respectively, and θ_p is the combined tolerance parameter. This extension adds to the computational overhead – since we now need two trees instead of one – but the extra work is compensated, to some extent, by the short range of the Pauli force, which permits a fairly large value of θ_p to be taken.

The three groups of potentials listed earlier (Skyrme/Yukawa, Coulomb, and Pauli) are approximated by their three corresponding pseudoparticle types. We therefore make three independent interaction lists for the force summation (subroutine RHS in the Frankfurt code). These lists are also used to compute the total potential energy. This requires some restructuring of the main DO loops in these routines. In the original code, there is a double loop over the particles:

```
do i = 1, np
    do k = i+1, np
        fx(i) = fx(i) + fik
        fx(k) = fx(k) − fik
        . . .
    end do
end do
```

The inner loop is replaced by three loops over each interaction list for particle i:

```
do i = 1, np
    do j = 1, nlist_sky_yuk(i)
        k = ilist_sky_yuk(i,j)
        fx(i) = fx(i) + fik
    end do

    do j = 1, nlist_coulomb(i)
        . . .
    end do
    do j = 1, nlist_pauli(i)
        . . .
    end do
end do
```

The partial forces f_{ik} now represent the particle–pseudoparticle interaction. In the process of restructuring, a number of inefficiencies have crept into the new force routine:

- The symmetry relation $f_{ki} = -f_{ik}$ exploited in the particle–particle (PP) scheme cannot be used in the tree version because we do not require the forces acting on the pseudoparticles (only those exerted by them on the particles).
- Each of the three pseudoparticle types has a unique position in space, which means that the separation distance r_{ik} must be evaluated in each inner loop. This effectively doubles the number of square-root operations per interaction pair compared with the PP scheme.
- The average vector length of the inner loops is lower than that in the PP scheme (e.g., 40 for $N = 400$, compared with 200 before). It can be increased by inverting the loop structure, so that the particle loop becomes the inner one. This requires a small amount of bookkeeping before each loop, but the net vector, scalar speed up for this routine is about 2.0, slightly less than the factor 2.5 obtained for the original PP code.

Besides the reduction in the number of force evaluations, the tree algorithm also permits an improvement in the treatment of the Yukawa force. In the original code, a problem arises because the Yukawa term contains a tabulated error function DERRF(X), which is limited to a maximum argument of $X = 6.5$. When the separation between particles increases such that this argument exceeds 6.5, their contribution is ignored. This results in poor energy conservation if it occurs *between* the Runge–Kutta steps, so particle pairs are flagged in the second step whenever they drift beyond this range. The whole force calculation is then repeated excluding these pairs, even if only a *single* pair is flagged.

In TREEQMD, this repetition is preempted by using the tree structure to flag particle–pseudoparticle pairs which *might* cross the Yukawa range (e.g., those for which the error function argument $X > 6.0$), and removing them from the Yukawa interaction in each Runge–Kutta step. This premature truncation does not appear to affect energy conservation, so the original problem is sidestepped.

Up to now, we have discussed the modifications to the QMD code which are necessary in order to evaluate the force and potential energy by the tree method. These consist of six tree-building routines: TREGRO3, CHARGE3, LIST3, TREGRO6, CHARGE6, and LIST6, which perform the tree construction, pseudoparticle definitions, and generation of interaction lists in 3D and 6D, respectively. These and the new force and energy routines are combined with the existing Runge–Kutta scheme as follows:

1st RK Step

- Build tree;
- evaluate forces.

2nd RK Step

- Redefine leaves and pseudoparticles using the *same tree as in Step 1*;
- evaluate forces using the *same interaction lists as in Step 1*;
- update positions and momenta.

The tree is not rebuilt for the 2nd Runge–Kutta step so that well-separated particle–pseudoparticle pairs removed from the Yukawa interaction in the 1st step are ignored in the 2nd step too. (If we rebuild the tree from scratch, we would lose this information, because the pair in question might no longer exist!) It is straightforward to extend the algorithm to 4th order. The tree is built afresh for the 1st step, and this structure is kept for the remaining three steps. In practice, we can get away with rebuilding the tree even less often (e.g., every five timesteps), assuming that the relationship between nearest neighbours does not change too quickly. Doing this virtually eliminates the tree-building overhead, which otherwise takes up about 30% of the cycle time. A similar approach of

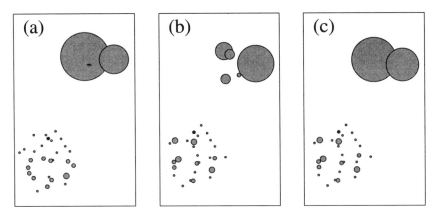

Fig. 3.5. Comparison of interaction lists at (a) $t = 0$, (b) $t = 5\Delta t$ for a tree rebuilt each timestep, and (c) $t = 5\Delta t$ for a tree structure frozen to that at $t = 0$.

adjusting rather than rebuilding the tree has also been used for astrophysical applications by McMillan and Aarseth (1993) and for studying planetesimal dynamics by Richardson (1993).

The effect of 'freezing' the tree structure is illustrated in Fig. 3.5, which shows the Yukawa interaction list for one of the target nuclei (marked X) at $t = 0$ and $t = 5\Delta t$, with and without rebuilding. One sees that the final lists are almost identical. The additional force error is negligible compared with that incurred by using a finite θ in the first place.

The first and simplest check of TREEQMD is to see how well the potential energy is reproduced for various tolerance criteria θ. In the limit of $\theta = 0$, we recover the result of direct particle–particle summation. Table 3.1 shows the potential energies for a randomly initialised Au+Au system, and the errors for each potential as calculated by TREEQMD. The same θ was used for the Pauli interaction as for the other 3D potentials.

We notice straight away that the Pauli potential is reproduced more accurately than the others. This is because its 6D interaction list is longer than the 3D lists, implying that we can use a larger opening angle θ for this interaction. The relatively large errors for the Skyrme potential stem from the rather crude approximation made earlier, namely that the pseudoparticle has the same width α as the individual particles. These two potentials depend quite strongly on the width as well as the separation, hence we see that they are reproduced less accurately than the others. The approximation is deficient because neighbouring wave packets have a combined extent larger than α. However, it is not obvious what we should actually take for the pseudopacket width. The underlying

Table 3.1. *Accuracy of energy evaluated by tree code for various* θ

Potential	Energy/nucleon	% Error for tree code			
	(MeV)	$\theta = 0.3$	$\theta = 0.5$	$\theta = 1.0$	$\theta = 1.5$
Coulomb	7.87	0.09	0.1	0.7	2.0
Pauli	15.08	0.05	0.06	0.15	1.5
Yukawa	-46.86	0.01	0.6	3.2	5.0
Skyrme-2	-23.78	0.13	0.8	6.0	9.0
Skyrme-3	33.94	0.2	1.0	8.0	13.0
Total P.E.	-13.74	0.04	0.9	2.1	5.7

physics suggests an average of the rms spatial spread and the packet widths of the daughter particles (just as the radius of the whole nucleus is calculated), but this would require a formalism similar to that used to treat 'variable-width' wave packets. Therefore, rather than attempt to make a width correction at this stage, we proceed with this approximation, and bear this in mind as a possible source of discrepancy between the TREEQMD and QMD codes.

The 'static' errors are an indication of how the tree approximation converges to the PP force-law as $\theta \to 0$; they do not necessarily predict how well the code will *conserve* energy (or reproduce other physical observables) when the particle trajectories are integrated in time. For this purpose, we use a $^{40}_{20}$Ca $+ \, ^{40}_{20}$Ca target-projectile system, initialised using a Monte Carlo minimisation of the total potential energy. The following quantities are monitored: the total energy (Fig. 3.6), and the momentum 'sphericity' of the system, defined as: $S \equiv 2P_z^2 - P_x^2 - P_y^2$ (see Fig. 3.7).

The error in total energy is defined by $\Delta E / E \equiv (E_{max} - E_{min})/E_{ave}$. For this example the error was 5% for $\theta = 1$ and 2% for $\theta = 0.5$, compared with 0.05% for the PP code.

The plot of sphericity in Fig. 3.7 indicates that both the $\theta = 0.5$ and $\theta = 1.0$ results follow the PP result up to the point of closest approach, but the solutions diverge thereafter, albeit more in phase than magnitude.

A more obvious physical difference between these three runs can be seen in Fig. 3.8, which shows the paths of the projectile and target centroids (in the $x - t$ plane). Prior to the collision, the trajectories obtained with the tree code lie inside those from the PP code, suggesting that there is an additional attractive force acting on the nucleons when the nuclei pass each other.

The benefit of using the tree algorithm in place of direct summation is fully realised only for heavy nuclei systems. For the target-projectile scenario there

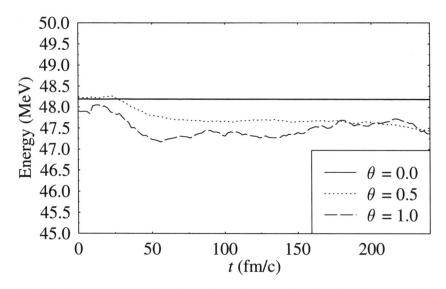

Fig. 3.6. Energy conservation of the Ca + Ca system for $\theta = 0$ (solid), $\theta = 0.5$ (dotted), and $\theta = 1.0$ (dashed).

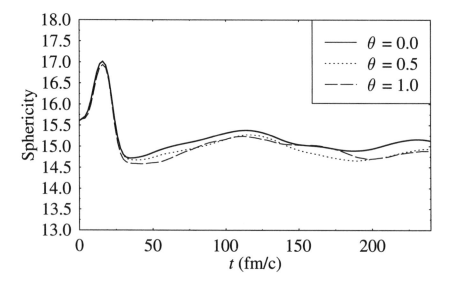

Fig. 3.7. Sphericity versus time for $\theta = 0$ (solid), $\theta = 0.5$ (dotted), and $\theta = 1.0$ (dashed).

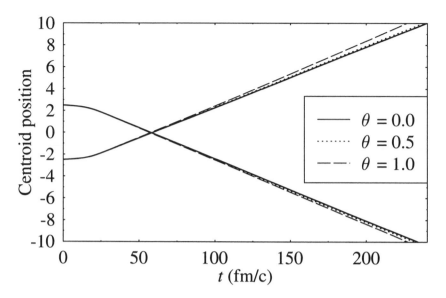

Fig. 3.8. *X*-component of centroid versus time of both target and projectile for $\theta = 0$ (solid), $\theta = 0.5$ (dotted), and $\theta = 1.0$ (dashed).

Table 3.2. *Timings of QMD vs TREEQMD for different θ.*

N	T_{qmd}/s	$\theta = 0.5$		$\theta = 1.0$	
		T_{tree}/s	Speedup	T_{tree}/s	Speedup
50	0.048	0.06	0.8	0.04	1.5
100	0.16	0.15	1.1	0.09	1.8
200	0.60	0.48	1.25	0.25	2.4
400	2.34	1.55	1.5	0.72	3.2
500	3.7	2.18	1.7	1.0	3.7

is a maximum of around 500 possible nucleons, although a more commonly studied event is Au + Au, with $N \simeq 400$. Table 3.2 gives a timing comparison between the vectorised version of QMD (Reusch 1992) and TREEQMD for two separate tolerance criteria θ. The timings were averaged only over the first ten timesteps to avoid any repetitions of the Runge–Kutta routine in QMD. This provides a measure of the basic cycle time.

In practice, the speedup for a whole event simulation is greater, because the QMD version often has to do additional work repeating the Runge–Kutta loop for the reason stated earlier. For example, an Au + Au event over a time

$200\,fm/c$ took 2,440s with QMD, compared to 540s with TREEQMD – an overall speedup of 4.5 in this case. The energy conservation in each case was 1.5% and 7.5%, respectively. The relatively poor conservation for the tree code was due in part to the rather large timestep used (reflected in the imperfect conservation by direct summation). Jerkiness in the particle motions caused by a large timestep tends to be amplified by the tree code.

In practice, the actual speedup of the new code (TREEQMD) over the original vectorised version depends on the system size and the desired accuracy of physical observables (such as the total energy and particle phase space). Speedups of 3–5 are comfortably achieved for heavy systems such as Au + Au, with energy conserved to better than 5%. The level of accuracy attained (compared with the old version using direct particle–particle summation) is surprisingly good considering the crudeness of the approximation used for the pseudoparticles in the tree code: namely, that a distant group of nucleon wave packets can be replaced by a single one of the same width and with an amplitude equal to the number of nucleons in the group. Ideally, the pseudoparticle should be a convolution of its constituent packets, but this would require a costly functional fit to the density distribution. A good compromise would possibly be to create a new Gaussian packet with a width comparable to the rms radius of the nucleon group, and an amplitude chosen such that the 'nucleon number' is conserved.

Whether such refinements are indeed necessary will depend on how the simulation results are ultimately used. One of the main goals of QMD is to model the formation of fragments – correlated groups of nucleons that bunch together after the collision. Analysis of this process requires repetition of the simulation many times (e.g., 1,000) with slightly different initial conditions to provide the necessary statistics for comparison with experiment. Small deviations in individual particle trajectories may have negligible influence on the final outcome, that is, the distribution of fragment charge and mass states.

A major difficulty in applying the tree method to QMD is to reduce the additional overheads to a minimum. This requirement is dictated by the fact that the number of nucleons in a two-nuclei collision is less than 500. Using a monopole approximation for clusters of pseudonucleons, the method is 'only' five times faster than the direct sum for these binary systems. However, TREEQMD might also be used to study much larger systems – such as infinite nuclear matter – in which several *thousand* nucleons are used in a simulation. For such applications, the tree method would be at least an order of magnitude faster, and would thus present new possibilities in computational nuclear physics theory.

4

Optimisation of Hierarchical Tree Codes

In Chapter 2 we saw the basic workings of the tree algorithm. Now we will discuss some methods that can be used to optimise the performance of this type of code. Although most of these techniques are not specific to tree codes, they are not always straightforward to implement within a hierarchical data structure. It therefore seems worthwhile to reconsider some of the common tricks of the N-body trade, in order to make sure that the tree code is optimised in every sense – not just in its $N \log N$ scaling.

There are basically two points of possible improvement:

- Improvement of the accuracy of the particle trajectory calculation by means of higher order integration schemes and individual timesteps. This is especially important for problems involving many close encounters of the particles, that is, 'collisional' problems.
- Speedup of the computation time needed to evaluate a single timestep by use of modern software and hardware combinations, such as vectorisation, and special-purpose hierarchical or parallel computer architecture.

4.1 Individual Timesteps

For most many-body simulations one would like the total simulated time $T = n_t \Delta t$ (where n_t is the number of timesteps) to be as large as possible to approach the hydrodynamic limit. However, the choice of the timestep Δt has to be a compromise between this aim and the fact that as Δt increases, the accuracy of the numerical integration gets rapidly worse. This becomes especially dramatic in cases where close encounters of the particles occur, and physically meaningful simulations will be obtained only over the total simulation time T, if the accumulated error is insignificant.

65

How does one choose an adequate timestep Δt for a given problem? In practice, Δt has to be chosen so that a good conservation of the total energy of the system can be achieved. One important parameter is the average collision frequency $\bar{\nu}$ of the system, which can be estimated to be

$$\bar{\nu} = \frac{v_{thermal}}{\lambda},$$

where $v_{thermal}$ is the root mean square of the velocities $v_{thermal} = \langle v^2 \rangle^{1/2}$ and λ is the average mean free path of the particles of the system. This mean free path is given by

$$\lambda = \frac{1}{n\sigma},$$

where n is the particle density and σ is the cross section for a collision. The precise definition of σ depends on the application of interest: For infrequent binary encounters one can just take the actual or estimated minimum distance of approach b_o (or 'impact parameter'), and set $\sigma = \pi b_o^2$. For plasma or particle beam applications, one usually defines the collision frequency as the inverse of the time taken to deflect a particle by 90°, which involves an averaging over many small-angle encounters (Kruer 1988). The timestep Δt of the simulation has to be smaller than the inverse of this collision frequency $\bar{\nu}$ and is therefore chosen according to

$$\Delta t_1 = \alpha \frac{n\sigma}{\langle v^2 \rangle^{1/2}}, \qquad (4.1)$$

where α is a free parameter.

However, if one is dealing with a collisional system a small number of particles of the system undergo close encounters. For these particles the potential and the force are changing very rapidly as they go along their path and a much smaller timestep is needed. Here, the relevant parameter is $\min(r_{ij})$, the distance between the ith particle and its nearest neighbour j. The path of the particle is simply $s = \frac{1}{2}\frac{F}{m}t^2$. The timestep is given by

$$\Delta t_2 = \alpha \sqrt{2 \min(r_{ij})\frac{m}{F}},$$

with the maximum allowed path being a fraction of $\min(r_{ij})$. If one is dealing with a force like $F = \beta/r^2$, the timestep is given by

$$\Delta t_2 = \alpha_2 [\min(r_{ij})]^{3/2} \qquad (4.2)$$

with $\alpha_2 = \alpha(2m/\beta)$. In cases where one of the particles in an encounter has a very high velocity the timestep given by (4.2) might still be too big. In this case, the adequate timestep can be obtained from the ratio of the distance to the nearest neighbour and the velocity v_{ij} of the fastest particle.

$$\Delta t_3 = \alpha \left[\min \frac{r_{ij}}{v_{ij}} \right]. \tag{4.3}$$

To ensure that the timestep is sufficiently small to evaluate the path of all particles accurately and to obtain a sufficient energy conservation, one should take the smallest of the timesteps given by (4.1)–(4.3)

$$\Delta t = \min(\Delta t_1, \Delta t_2, \Delta t_3). \tag{4.4}$$

Taking such a short timestep is often wasteful because only a very limited number of particles are actually involved in close encounters or have relatively high velocities; the majority of particles proceed rather undisturbed and would not need such a small timestep. One way to reduce the computation time is to recalculate the paths of the particles in close encounters more often than those of the rest of the system, therefore introducing two or more different timesteps. In conventional particle–particle codes this has been implemented even to the extent of taking variable and individual timesteps for each particle (Aarseth 1963, Aarseth & Hoyle 1964, Wielen 1967, Hayli 1967, Ahmad & Cohen 1973).

Different timescales could be introduced to tree codes quite easily too. Assuming one wants to perform the calculation on two different timescales – a small one Δt_c for the particles involved in close encounters and a larger one Δt_a for the rest of the particles of the system – one would have to sort out the particles which undergo a close encounter into a seperate list. The timestep Δt_a has to be chosen according to Eq. 4.1 and the timestep Δt_c as the smaller one of Eqs. 4.2 and 4.3. Moreover, it is preferable that the timesteps Δt_a to be a multiple of Δt_c,

$$\Delta t_a = n_t \Delta t_c, \tag{4.5}$$

in order to make it easy to include all particles in the calculation after n_t timesteps.

For the integration on the smaller timescale, only the force on the particles involved in close encounters is evaluated. However, in order to do this one has to update the positions of all particles as well as build the tree structure afresh each time. As we have seen in Chapter 2, building the tree structure takes only a fraction of the time the force calculation takes, so the extra overhead involved is small. After n_t timesteps Δt_c the force on all particles is calculated and a new list of the particles which need a smaller timestep is created. Figure 4.1 shows a flowchart of a tree code with two different timescales. This technique

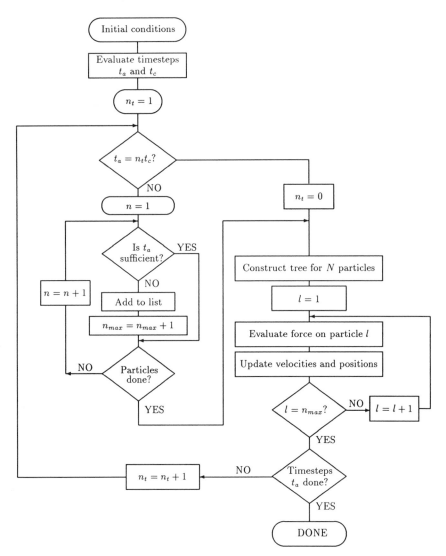

Fig. 4.1. Tree algorithm incorporating two timescales.

can in principle be extended to allow for more timescales, as illustrated in Fig. 4.2. but rebuilding the tree for each (smallest) timestep would eventually cancel the gain in efficiency of the force calculation. Nevertheless, multitimestep tree codes do exist: Implementations using a binary time-structure have been developed by Hernquist and Katz (1989) for 'Smooth Particle Hydrodynamics' (SPH) simulations of self-gravitating, fluid-like astrophysical objects, and by

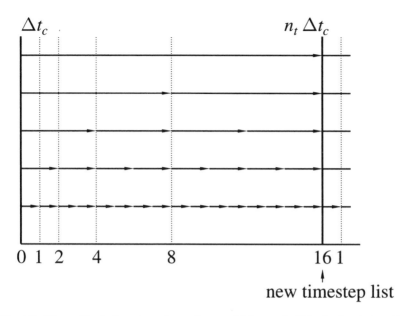

Fig. 4.2. Binary block-timestep scheme for a multitimescale N-body simulation. Particles are assigned timesteps of Δt_c, $2\Delta t_c$, $4\Delta t_c$, etc., according to their position and velocity.

McMillan and Aarseth (1993) for collisional stellar systems. This idea has also been extended to explicitly incorporate the pseudoparticles of the tree in the integration scheme to avoid rebuilding every timestep (Jernighan & Porter 1989).

Our simplified two-timescale algorithm can be taken a bit farther, and one can recalculate at each timestep t_c the smallest necessary timestep t_a for the present state of the system. This makes the algorithm more effective, on the one hand, because it avoids having an unnecessary small timestep in times where there are no close collisions in the system; on the other hand, it makes the calculation more accurate, because during the evolution of the system much closer encounters could develop than in the initial distribution of the particles.

Apart from the previously discussed method there have been different suggestions to solve the problem of close collisions. One way is to treat the close encounters of two particles (binaries) separately and make use of analytical solutions: A technique known as Kustannheimo–Stiefel regularisation (Kustannheimo & Stiefel 1965). This works only for systems where three (and more) body collisions are rare and the effect of the rest of the system can be ignored.

Another suggestion has been to create a tree code which uses an additional tree structure in time. This sounds like an elegant solution to the problem, but would make the code quite complex. To the best of our knowledge, such a code has not yet been worked out in detail.

4.2 Higher Order Integration Schemes

For a given value of the timestep Δt the accuracy of the simulation of the dynamics of the system can be improved by using higher order integration algorithms or predictor-corrector cycles. These algorithms generally achieve better energy conservation.

The most obvious way to obtain the new positions $\mathbf{r}_i(t + \Delta t)$ and velocities $\mathbf{v}_i(t + \Delta t)$ after one timestep Δt, given the old positions $\mathbf{r}_i(t)$ and velocities $\mathbf{v}_i(t)$, is to calculate first the velocities from the accelerations $\mathbf{a}_i(t) = \ddot{\mathbf{r}}_i(t)$. The accelerations are obtained directly from the forces $\mathbf{F}_i(t)$ evaluated by the tree algorithm. The velocities are

$$\mathbf{v}_i(t + \Delta t) = \mathbf{v}_i(t) + \mathbf{a}_i(t)\Delta t \tag{4.6}$$

and the positions are obtained by a Taylor expansion

$$\mathbf{r}_i(t + \Delta t) = \mathbf{r}_i(t) + \mathbf{v}_i(t)\Delta t + \frac{1}{2}\mathbf{a}_i(t)(\Delta t)^2. \tag{4.7}$$

Alternatively, one can calculate the positions first, as in the following algorithm introduced by Verlet (1967). Here the positions and velocities are given by

$$\mathbf{r}_i(t + \Delta t) = 2\mathbf{r}_i(t) - \mathbf{r}_i(t - \Delta t) + \mathbf{a}_i(t)(\Delta t)^2, \tag{4.8}$$

$$\mathbf{v}_i(t) = \frac{1}{2}\left[\mathbf{r}_i(t + \Delta t) - \mathbf{r}_i(t - \Delta t)\right]/\Delta t. \tag{4.9}$$

Not only are the positions at the present and the next timestep used but also the positions at the previous timestep (see Fig. 4.3). The error of $\mathbf{r}_i(t + \Delta t)$ is of the order $O((\Delta t)^4)$, whereas that of $\mathbf{v}_i(t + \Delta t)$ is of the order $O((\Delta t)^2)$. However, the relatively large error in the velocity calculation is not very important, because the velocities play no part in the integration of the equation of motion.

This algorithm is time-reversible, and provided the forces are conservative, it also guarantees conservation of linear momentum. Despite these favourable properties, along with good energy conservation, this scheme can in practice lead to numerical imprecision. This is because for small accelerations, Eq. 4.9 may introduce a rounding error by adding a small $O(\delta t^2)$ term to a difference of large $O(\Delta t^0)$ terms.

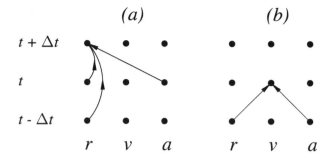

Fig. 4.3. Verlet integration scheme.

The preferred scheme, now widely adapted, is the so-called leap-frog algorithm. This avoids the threat of a rounding error by incorporating the velocities explicitly at the midstep of the particle positions:

$$\mathbf{v}_i(t + \frac{\Delta t}{2}) = \mathbf{v}_i(t - \frac{\Delta t}{2}) + \mathbf{a}_i(t)(\Delta t), \tag{4.10}$$

$$\mathbf{r}_i(t + \Delta t) = \mathbf{r}_i(t) + \mathbf{v}_i(t + \frac{\Delta t}{2})\Delta t. \tag{4.11}$$

Note how the velocities and positions now alternate (see Fig. 4.4). After each timestep, the velocities are always defined half a step earlier than the positions, so to check energy conservation, and so forth, we need to store:

$$\mathbf{v}_i(t) = \frac{1}{2}\left[\mathbf{v}_i(t + \frac{\Delta t}{2}) + \mathbf{v}_i(t - \frac{\Delta t}{2})\right].$$

Schofield (1973) and Beeman (1976) used an algorithm which shows a significantly improved energy conservation especially for the Lennard–Jones (12, 6) potentials. It is given by

$$\mathbf{r}_i(t + \Delta t) = \mathbf{r}_i(t) + \mathbf{v}_i(t)\Delta t + \frac{1}{6}\left[4\mathbf{a}_i(t) - \mathbf{a}_i(t - \Delta t)\right](\Delta t)^2$$

$$\mathbf{v}_i(t + \Delta t) = \mathbf{v}_i(t) + \frac{1}{6}\left[2\mathbf{a}_i(t + \Delta t) + 5\mathbf{a}_i(t) - \mathbf{a}_i(t - \Delta t)\Delta t\right]$$

and basically provides better treatment of the velocities (Sangster & Dixon 1976, Eastwood & Hockney 1974). For very steep gradients the leap-frog algorithm can become unstable; in this case a Two-Step Lax–Wendroff scheme can be used, which is second order in time, but defines intermediate values at the half timestep (Press et al. 1989). One way to obtain even higher accuracy is to include more terms in the Taylor expansion. There are several algorithms

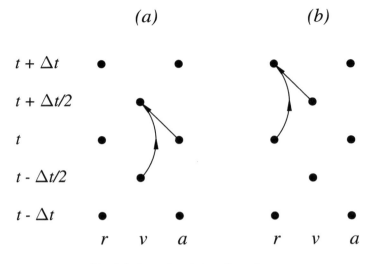

Fig. 4.4. Leap-frog integration scheme.

which make use of this, for example, the well-known Runge–Kutta method. For a description of the various methods see, for example, Press et al. (1989).

As an alternative to these methods based on the Taylor expansion one can use a predictor-corrector method (Rahman 1964). This algorithm is more suitable for short-range forces and starts with an estimate of the new positions $\mathbf{r}_i^e(t+\Delta t)$ and velocities $\mathbf{v}_i^e(t+\Delta t)$

$$\mathbf{r}_i^e(t+\Delta t) = \mathbf{r}_i(t) + \mathbf{v}_i(t)\Delta t \tag{4.12}$$

$$\mathbf{v}_i^e(t+\Delta t) = \mathbf{v}_i(t) + \frac{1}{2}[\mathbf{a}_i^e(t+\Delta t) + \mathbf{a}_i(t)]\Delta t, \tag{4.13}$$

where again $\mathbf{a}_i^e(t+\Delta t)$ is obtained from $\mathbf{r}_i^e(t+\Delta t)$ and the known force $\mathbf{F}_i(t)$. The corrected values of the positions are then given by

$$\mathbf{r}_i(t+\Delta t) = \mathbf{r}_i(t) + \frac{1}{2}[\mathbf{v}_i^e(t+\Delta t) + \mathbf{v}_i(t)]\Delta t. \tag{4.14}$$

Iterating about Eqs. 4.12–4.14, this procedure can be repeated until the required convergence is achieved.

In all of the algorithms described, whether based on a high-order Taylor expansion or a predictor-corrector method, one has some additional computational costs. One usually can use a larger timestep, but the updating of the positions and velocities takes longer. When implementing these schemes, the improvement in accuracy should be balanced with these additional costs. As a rule it is better to start with a simple, robust scheme such as leap-frog before attempting to implement something more refined that may be difficult to debug.

4.3 Vectorisation and Parallelisation

A common reaction to the suggestion that a problem could be solved faster by a tree code than by direct methods is: "Yes, but my force-summation routine vectorises so efficiently that I get a speedup of 10 on machine XYZ. Since you can't vectorise a hierarchical code structure, you may as well stick with the direct method." This objection is based on two misconceptions. First, tree codes *can* be vectorised – and efficiently too for a large enough particle number; second, vectorised or not, the CPU-time for the direct method remains $O(N^2)$, so sooner or later the tree code will catch up. There is no reason to suppose that the need for solutions to N-body problems will cease at some magical number of bodies. Certainly, there is a class of problems for which N is limited. For instance, a protein has a well-defined number of atoms, so as soon as one has a model that can manage the biggest protein of interest in an acceptable time, there is little scientific incentive to optimise further – though there may be a financial one, perhaps. However, there is also a large class of problems where N is far greater (e.g., 10^{23}) than any simulation will be able to cope with in the foreseeable future, but for which accuracy in representing reality depends on having N as large as possible to provide good statistics. By optimising the tree algorithm, we bring down the 'cross-over' point – the number of particles for which the hierarchical method becomes more efficient than the direct method. This can be anything from a few hundred to several thousand, depending on the complexity and the number of dimensions in the force law.

As we saw in Chapter 2, the tree is built in a recursive manner, starting with a box containing all the simulation particles, and repeatedly subdividing this region until each box contains a single particle. The easiest way to implement this in a program is to immediately subdivide the first 'subbox' (e.g., the top-left in Fig. 2.1) and continue the subdivision until the first leaf is reached, creating the twig-nodes along the way. One then moves back down the tree to find the nearest unprocessed twig, at which point the subdivision procedure can be repeated. This divide-up, search-down process is continued until there are no more twigs to subdivide, as illustrated schematically by Fig. 4.5.

As it stands, our tree-build algorithm lends itself perfectly to implementation in a language capable of recursive subroutine calls, such as C or PASCAL, but there is not much parallelism. We can deal with only one node at a time, so assuming we loop over the particles attached to that node, the vector length will decrease rapidly with increasing tree level.

1st step

2nd step

3rd step

4th step

5th step

6th step

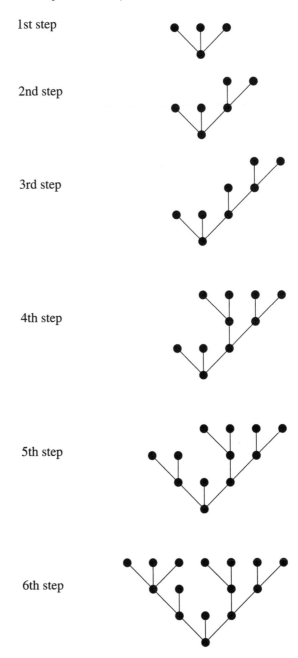

Fig. 4.5. Recursive tree-build.

An alternative to this node-by-node search is to subdivide all subboxes of the same level simultaneously, as shown in Fig. 4.6. The particle loop is modified so that it contains a *list* of particles not yet sitting in their own cell, that is, those sharing a cell with one or more other particles. In our tree terminology, the effective vector length of this loop is the total number of particles "still attached to twig-nodes", which, apart from the last few levels of the tree, is practically equal to N. The price to be paid for this improvement in efficiency is some additional bookkeeping. The vectorised tree-build algorithm in detail is as follows.

> **for** each particle p **do**:
>   attach to root node
>   add to list
>   label as *not_done*
> **end do**
>
> level = 1
> **while** any particle *not_done*:
>   **for** each particle in list **do**:
>   find out which subcell it is in
>   **end do**
>
>   count particles in each subcell
>
>   **for** each subcell **do**:
>   find out whether it is a leaf or twig
>   **end do**
>
>   **for** each new leaf **do**:
>   set particle label as *done*
>   **end do**
>
>   **for** each new twig **do**:
>   store twig node in tree
>   store twig in list of nodes for next level
>   **end do**
>
>   Rebuild particle list:
>   np_new = 0

A

```
|              for each particle in list do:
|                  if (not_done) then
|                      np_new = np_new + 1
B                      put particle in new list
|                      reattach to subcell twig node
|                  end if
|              end do
              level = level + 1
          end while
```

The loops A and B of the particle list make use of 'gather-scatter' operations, or indirect addressing, which are readily vectorised on nearly all modern supercomputers. The remaining loops over the subcells, leaves, and twigs are also vectorisable, but it is best to do these conditionally since their length can sometimes be small.

The completed tree is then postprocessed to obtain the physical quantities needed for the twig-nodes, such as the mass (charge), centre of mass (charge), and multipole moments. These are obtained, as in the recursive algorithm, by a downward sweep of the tree from the leaves to the root. Vectorisation is again made possible by looping over all nodes at the same level in the tree.

The tree construction and moments calculation typically take 3–10% of the total time. Like all N-body codes, most of the time used by a tree code is spent computing forces between the particles. It is therefore more important to optimise this routine than the tree-building part. Once the data structure is in place, however, it is a relatively straightforward matter to hop from node to node to obtain the desired sum. Vectorisation depends on choosing the right *order* of nodes during the tree traversal. Before we explain in detail how this is done, it is instructive to recall the recursive force summation routine described in Chapter 2, which we can summarise as follows.

```
          for each particle p do:
              attach to root node
              while (sum not finished):
                  if (node(p) far enough away or leaf) then
                      sum force: f(p) = f(p) + f_{p,node(p)}

                  else
                          resolve it into daughter nodes
                  end if
              end while
          end do
```

1st step

2nd step

3rd step

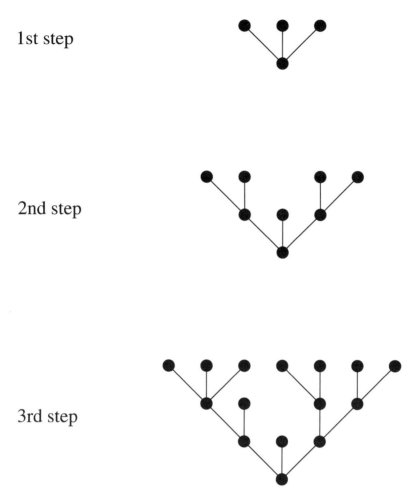

Fig. 4.6. Vectorised tree-build.

This algorithm basically comprises two loops: An outer one over the particles and an inner one over the interaction list for a given particle. A node is resolved when $s/d < \theta$, where as before, s is the cell size, d is the distance from the particle to the pseudoparticle (node) centre-of-mass, and θ is the tolerance criterion. As when building the tree, the 'resolve into daughter nodes' instruction is the main obstacle to vectorisation, and ensures that a lot of time is spent in tree traversals (see Fig. 4.7).

The optimal way of 'unravelling' the particle and interaction loops depends somewhat on the machine architecture and the application in question. Three

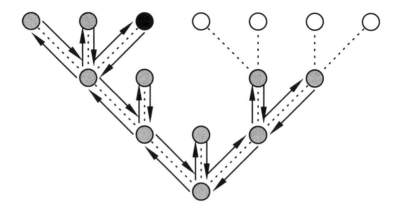

Fig. 4.7. Recursive tree search: interaction list (grey nodes) is generated for a black particle.

such alternatives were developed in Princeton and published simultaneously in 1990. The first of these, by Hernquist (1990), vectorises tree searches over all nodes at the same level in a similar manner to the tree-building algorithm described previously. By contrast, the method implemented by Makino (1990b) vectorises the tree search across particles, using additional bookkeeping to perform the tree traversals for each particle independently. Finally, Barnes exploited the fact that the 'interaction list' – the list of particles and clusters of particles which make up the force sum – is very similar for near neighbours (Barnes 1990), thus shifting the bulk of the effort from tree searches to the readily vectorisable force summation itself. Which of these methods is better depends on the machine the code is running on, but it is worth noting that the Barnes and Hernquist algorithms can be combined, resulting in an algorithm which is faster than Makino's by itself.

The vectorised algorithm presented here largely follows Makino's, for no better reason than that we implemented this ourselves (Gibbon 1992b). However, instead of summing the force immediately, we first determine the interaction lists: `node_list` (p,i) as follows.

	attach to root
|	**do** p = 1, n
A	done(p) = .false.
|	node(p) = 0
|	**end do**

while (any particle *not_done*) **do**:

| **do** p = 1, n
| **if** (node(p) far away from p **or** node(p) = leaf)
| add(p) = .true.
| node(p) = next(node(p))
B   **else**
| add(p) = .false.
| node(p) = daughter(node(p))
| **endif**
| **if** (node(p) = 0) done(p) = .true.
| **end do**

Select terms to be added to interaction lists:

    n_add = 0
| **do** p = 1, n
| **if** (add(p) = .true.) **then**
| n_add = n_add + 1
C   p_add(n_add) = p
| **endif**
| **end do**

Update interaction lists:

| **do** i = 1, n_add
| p = p_add(i)
D   nlist(p) = nlist(p) + 1
| node_list(p, nlist(p)) = node(p)
| **end do**

    end while

This is illustrated schematically in Fig. 4.8. We see immediately that there are fewer traversals between nodes compared with Fig. 4.7.

The main difference between this algorithm and Makino's is that we have divided the procedure for the interaction list update into three loops B, C, and D: Loop B tests whether to accept or resolve the particles' current node (i.e., checks the condition $s/d < \theta$ for the relevant particle-node pair); loop C makes

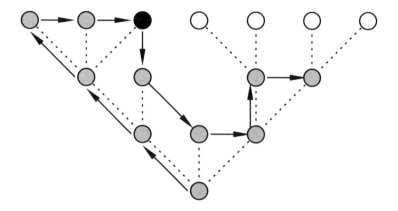

Fig. 4.8. Vectorised tree-search: interaction list generated for black particle.

a *reduced* list of particles which have nodes to be added; loop D updates the interaction lists. This has two advantages. First, we avoid wasting 'vector idling' time in the **IF** block of loop B caused by not adding a node. As it is, the two branches are exactly balanced: If, instead, we inserted the force sum or updated the interaction list at this point the 'add' branch would take much longer than the 'subdivide' branch. This means that the effective vector length of the loop would be reduced to $O(\text{n\_add})$, which is generally much less than N. Second, we can further sort the added nodes into twigs, leaves – and if necessary, charge sign – which saves time when a multipole expansion is used for the force and potential.

In practice, a further loop is added to remove the particles that have completed the tree-walk. Thus, p is replaced by an additional list particle_list(p) which is updated after loop D as follows.

```
n_new = 0
do i = 1, np
    if (done(i) = .false.) then
        n_new = n_new + 1
        particle_list (n_new) = particle_list(i)
        node (n_new) = node(i)
        done (n_new) = done(i)
    endif
end do
np = n_new
```

Finally, the forces are summed by a simple double-loop:

```
do p = 1, n
    f(p) = 0
    do i = 1, nlist(p)
        f(p) = f(p) + f_{p,node_list(i)}
    end do
end do
```

4.4 Timing

We now turn to the problem of quantifying the optimisations discussed previously: How well does a tree code actually perform? Because of the tradeoff between speed and accuracy described earlier in Section 2.2, this question is not as straightforward to answer as it might seem. Some extensive analyses of tree-code performance in the astrophysical context have been made by Hernquist (1987, 1990), Barnes (1990), and Makino (1990b).

In general, the time required to sum the forces using the tree structure depends on a number of factors: The tolerance parameter θ, the number of terms in the multipole expansion, and the density distribution. The simplest case to consider is the monopole tree code typically used for gravitional problems. Figure 4.9 compares the time taken to complete the force summation for various values of θ, together with the time for the direct PP algorithm. In this example, and the ones that follow, we have left the time in arbitrary units, because the absolute timing is machine-dependent.

Two things are apparent from this plot: (i) the transition in scaling from $O(N^2)$ to $O(N)$ as θ is increased; (ii) the point at which the tree code becomes more efficient than PP is reached sooner as θ is increased.

If we add more terms to the multipole expansion, then we increase the 'overhead' of the tree code. This is illustrated in Fig. 4.10, which shows a comparison between PP and $\theta = 0.5$ with monopole and quadrupole terms included, respectively.

Here we see that the slope of the tree-code curves is the same, but the overhead is multiplied by two when quadrupole terms are added, reflecting the increased number of operations in the force expansion. Comparing Fig. 4.10 with Fig. 4.9, we notice that for $N < 10^3$, it would actually be more efficient to use a monopole code with $\theta = 0.3$ than a quadrupole version with $\theta = 0.5$. The eventual choice of parameters depends on the problem at hand – if accuracy is important, then one may be better off with the higher order expansion.

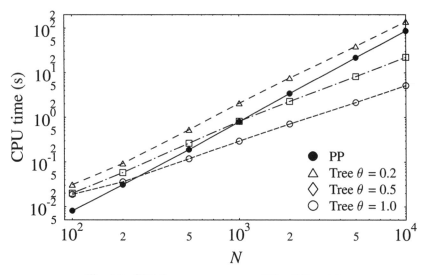

Fig. 4.9. CPU time per step versus N for different θ.

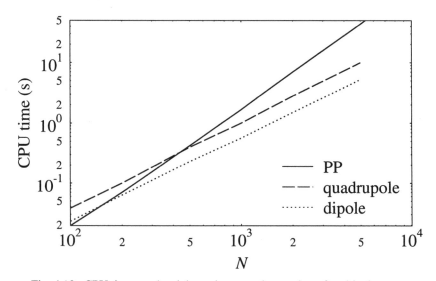

Fig. 4.10. CPU time overhead dependence on the number of multipole terms.

4.5 Accuracy

This brings us to another issue: how to measure the accuracy of a tree code, or for that matter, any algorithm that makes use of some approximation in the force or potential energy calculation. It is important here to distinguish between absolute accuracy, for example, how many significant figures to which the force on one particle can be computed – and how well a given global variable is *conserved* in time. High precision in the force summation of the tree code is often but not always the limiting factor for good energy conservation. On the other hand, good energy conservation does not imply good momentum conservation, as we noted in Chapter 2.

Definitions of the force error vary in the literature, but a commonly used quantity is:

$$\epsilon_k = \left\{ \frac{\sum_{i=1}^{N_{test}} \left(F_{ki}^{tree} - F_{ki}^{direct} \right)^2}{\sum_{i=1}^{N_{test}} \left(F_{ki}^{tree} \right)^2} \right\}^{\frac{1}{2}}, \qquad (4.15)$$

where F_{ki}^{tree} and F_{ki}^{direct} are the kth components of the forces on particle i, evaluated using the tree and direct (PP) methods, respectively. N_{test} is an arbitrary number of randomly chosen test particles, typically much smaller than N. One can generally take the average over the components x, y, and z.

Similarly, for the potential we can define:

$$\epsilon_0 = \left\{ \frac{\sum_{i=1}^{N_{test}} \left(\Phi_i^{tree} - \Phi_i^{direct} \right)^2}{\sum_{i=1}^{N_{test}} \left(\Phi_i^{tree} \right)^2} \right\}^{\frac{1}{2}}, \qquad (4.16)$$

where Φ_i^{tree} and Φ_i^{direct} are the potentials for particle i, evaluated with the tree and PP methods, respectively.

We can use (4.15) or (4.16) to illustrate the tradeoff between speed and accuracy characteristic of hierarchical methods. Figure 4.11 shows the variation of CPU time and error as a function of θ. We see that for $\theta > 1$, we begin to sacrifice a lot of accuracy for little gain in speed. In principle, one can improve on this by adding terms in the multipole expansion – as seen in Fig. 4.12, which shows the force errors obtained from a Plummer model calculation (McMillan & Aarseth 1993).

On the other hand, what if we keep increasing θ until there is only one pseudoparticle – i.e., the root cell – in each interaction list? We would have to exclude the particles and its immediate neighbours from this term so that the multipole expansion stays valid. The algorithm would then have a CPU scaling of $O(N)$, and provided we keep enough multipole terms, it should eventually outperform the standard tree code. This is essentially the idea behind the Fast

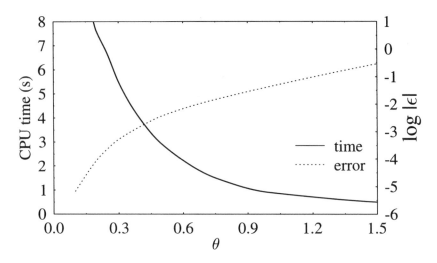

Fig. 4.11. Tradeoff between CPU time per step and average force error for the tree code with monopole terms only.

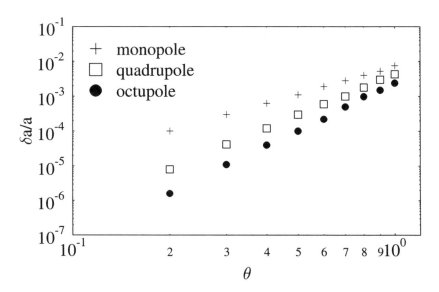

Fig. 4.12. Median force errors for $N = 1,024$ (small symbols) and 4,096 (larger symbols). The monopole, quadrupole, and octupole errors scale as $\theta^{2.7}$, $\theta^{4.0}$, and $\theta^{4.6}$, respectively. Adapted from McMillan and Aarseth (1993), courtesy of S. McMillan. ©1993. The American Astronomical Society (University of Chicago Press).

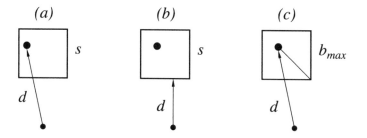

Fig. 4.13. Multipole acceptance criteria: (a) Barnes–Hut, (b) minimum-distance and (c) B_{max} ($b_{max}/d < \theta$).

Multipole Method (FMM) developed by Greengard (1988), except that neighbouring particles are still summed directly, and intermediate 'local' expansions are used to carry the multipole moments from the root cell to the particle positions. Although the FMM can be regarded as a refinement of the tree algorithm in this sense, it was actually developed independently in an entirely different field. The code structure is also quite different and for this reason we describe the FMM separately in Chapter 7.

Apart from the FMM, there are also other ways to improve the accuracy of a standard tree code. In particular, the acceptance criterion (2.4) introduced in Chapter 2 can be replaced by more sophisticated tests which guarantee an upper error bound for the multipole approximation, similar to those derived by Greengard (1988) for FMM. Motivated by a wish to apply the same mathematical rigour which underpins FMM to tree codes, Salmon and Warren (1994) suggested a number of alternative 'multipole acceptance criteria' (MACs) which avoid large errors in certain unusual situations.

One such possibility is to use replace $d = |\mathbf{r}_p - \mathbf{r}_{cm}|$ with $d = \min |\mathbf{r}_p - \mathbf{r}_{edge}|$, where $\mathbf{r}_{edge}$ is the position of the edge of the cell. Another is to replace the size of the cell s with the maximum distance from the edge of the cell to the centre of mass $b_{max} = \max |\mathbf{r}_{cm} - \mathbf{r}_{edge}|$. These criteria are illustrated in Fig. 4.13. These two alternatives, the 'minimum-distance' and 'B_{max}' criteria, respectively, have the advantage that they avoid catastrophic errors for configurations where a large mass is concentrated at the corner of a cell. Apart from that, they give a similar performance for a given accuracy level.

Salmon and Warren also describe a more systematic but involved approach in which a given accuracy in the force or potential can be *guaranteed*. To do this, they derive a worst-case error for the acceleration after adding a p-term multipole expansion in the force-summation:

$$\Delta a \leq \frac{1}{d^2} \frac{1}{(1 - b_{max}/d)^2}$$

$$\times \left[(p+2)\frac{|M_{p+1}|^{\text{upper}}}{d^{p+1}} - (p+1)\frac{|M_{p+2}|^{\text{lower}}}{d^{p+2}} \right]$$

$$(4.17)$$

where M_p is the pth multipole moment and $|X|^{\text{upper}}$ and $|X|^{\text{lower}}$ denote upper and lower bounds of a quantity X, respectively. The simple geometrical criterion is then replaced by:

$$\Delta a_{cell} \leq \Delta_{max},$$

where $|\Delta a_{cell}|^{\text{upper}}$ is the maximum acceleration error of a node of the tree given by the RHS of (4.17), and Δ_{max} is a user-specified tolerance parameter. This can be developed further, for instance by preprocessing the tree to store values of d for which the criterion is satisfied, and then obtaining a mean or rms absolute error, though this then requires some modifications to the standard tree code. However, preliminary tests by Salmon and Warren suggest that this approach ultimately provides a better performance than the simple MACs.

For dynamic applications, one is often less concerned with the force accuracy as with energy or momentum conservation. The effect of hierarchical clustering on these laws has been documented extensively by Hernquist (1987) for simple astrophysical systems. In a PP code, energy is conserved exactly as $\Delta t \to 0$. This is *not* true of a tree code, however. For a given Δt, and typical $\theta < 1$, the nonconservation of energy due to truncation error in the integration scheme will be comparable to the error from the force calculation. Thus, as $\Delta t \to 0$, there will always remain an error due to clustering.

From another point of view, for typical timesteps the energy conservation for PP and tree codes will look rather similar. Moreover, there will be some value of θ below which energy conservation of a tree code will not improve. One should note, however, that the behaviour may be somewhat different for periodic systems, in which additional errors may arise from the combination of hierarchical clustering and image particles; see, for example, Bouchet and Hernquist (1988), and Hernquist, Bouchet, and Suto (1991).

The improvements in speed gained by vectorisation are very much algorithm- and machine-dependent, as discussed earlier in Section 4.3. One feature common to many implementations, however, is that the speedup of a vectorised code over a scalar one tends to increase with N. This is because for a typical

θ (0.3–1.0), the average length of interaction lists only becomes comparable with the vector length of the hardware for $N > 1,000$–$5,000$. In practice one can expect an overall speedup anywhere between 2 and 40 depending on the machine, but generally less than that obtained with a PP code. Details of speed improvement on different machines can be found in Hernquist (1990), Barnes (1990), Makino (1990b), and Suginohara et al. (1991).

4.6 Special Hardware

Though not strictly in the spirit of a book about fast N-body *methods*, we include here a brief mention of another approach to many-body simulations which relies on sheer processing power rather than algorithmic finesse. One such example is the 'digital Orrery' (Applegate et al. 1985), originally designed for orbital mechanics problems. This uses n identical hardware units, or 'planet computers' to represent the simulation bodies. This machine, based on an SIMD (Single Instruction, Multiple Data) architecture, allows forces to be computed in $O(n)$ time by accumulating information around a ring between the planet computers. A central processor then collects the results from each planet and updates the equation of motion. As for other parallel architectures, this technique can be extended to other problems by assigning N/n particles to each processor.

A similar approach has been adopted by Sugimoto and collaborators (1990, 1993), who have built a special-purpose processor ('GRAPE', for gravity pipe) to perform the force-summation for each particle via a 'pipeline'. This uses chips with hard-wired functions such as $f^{-1.5}$ and x^2 to achieve gigaflop performance at a fraction of the hardware cost of comparable machines. Work is currently under way to extend this technique using parallel pipelines to reach teraflop performance in the near future.

As far as tree codes are concerned, it is not immediately obvious that such special hardware would be advantageous, because the additional effort needed to unravel and extract information from the tree might degrade the performance somewhat. Nevertheless, as McMillan and Aarseth point out (McMillan & Aarseth 1993), even order-of-magnitude improvements in hardware speed increase the maximum manageable number of bodies by a factor of only 2.0 or so: A hybrid approach combining the Grape architecture with a tree code may offer the best of both worlds.

5

Periodic Boundary Conditions

The gridlessness of the tree code lends itself naturally to open boundaries, in which the simulation region adjusts to the spatial distribution of the particles. While this is an obvious way to treat problems of the kind described in Chapter 3 – such as colliding galaxies and propagating beams – a large class of N-body problems can be tackled only by modelling a small region of the macroscopic object by means of periodic boundaries. In order to eliminate surface effects, an infinite system is constructed from identical cells containing N particles each (see Fig. 5.1).

The number of particles in the cell has to be constant throughout the simulation; this can be ensured through the use of 'penetrable' walls. Due to the periodicity, this means that as a particle leaves the box through one wall, it re-enters with unchanged velocity through the opposite wall. It can be shown that for gases the difference in the virial coefficients between an infinite system and a system consisting of a finite number of N molecules with periodic boundaries is of the order $1/N$ (Alder & Wainwright 1959).

Assuming that N particles interact through Coulomb forces the total energy including the interaction with image particles is given by

$$V_c = \frac{1}{2} \sum_{\mathbf{n}} \sum_{i=1}^{N} \sum_{j=1}^{N} \frac{Z_i Z_j}{r_{\mathbf{n},i,j}}, \tag{5.1}$$

where the sum over the vector $\mathbf{n}$ represents the image boxes. An equivalent expression exists for gravitational forces. Because the potential is long range, this sum converges very slowly, and would necessitate huge amounts of computing time. Fortunately, there are two alternatives to evaluating Eq. 5.1 as it stands:

- Minimum image method
- Ewald summation

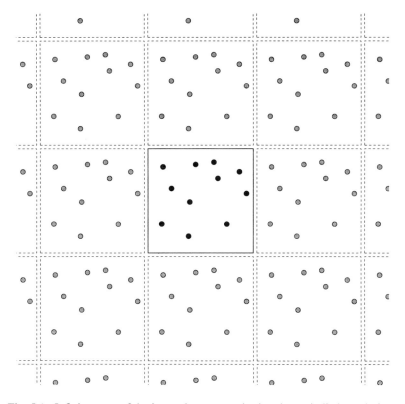

Fig. 5.1. Infinite array of the image boxes constituting the periodic boundaries.

In the minimum image method the particles are allowed to interact only with other particles in the basic cell or one of the directly neighbouring cells. Each individual particle interacts only with the $N - 1$ particles that happen to be located in a cube of length L centred at the position of this particle (see Fig. 5.2). As in the fully periodic case, particles exiting the simulation volume re-enter from the opposite side. While this is sufficient for weakly coupled systems where $\Gamma \leq 1$, and the Coulomb potential is shielded over distances longer than a Debye sphere, it introduces severe errors in strongly coupled systems, where $\Gamma \gg 1$. In this case, each particle must be allowed to interact with all the periodic images of all the other particles, as implied by Eq. 5.1. In the Ewald transformation this slowly converging sum is replaced by two rapidly convergent series, thus including the leading terms of all the image particles in the calculation. We now discuss in more detail these two different methods for dealing with periodic boundaries and their implementation in tree codes.

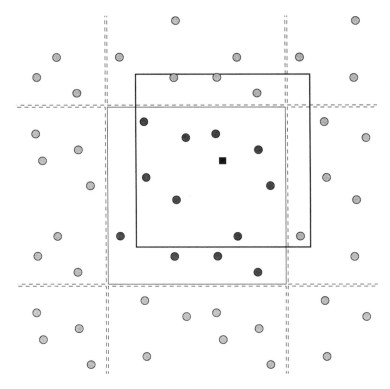

Fig. 5.2. Minimum image array.

5.1 Minimum Image Method

The minimum image method cuts out a cube of length L centred at the particle of consideration from the array of periodic images, as shown in Fig. 5.2, so that only the influence of particles within this volume are taken into account. For PP codes this procedure is straightforward, but for tree codes it is not so simple. It would be much too time consuming to rebuild the tree for each particle, but the alternative of retaining the interaction list for the original box presents another problem. When placing the centric cube around the particle, most of the boundary pseudoparticles get bisected, and end up contributing partly to the cube and partly to the outside region. This is demonstrated in the 2D example of Fig. 5.3.

Instead of subdividing all cells straddling the boundaries, which would mean many more interaction terms, Bouchet and Hernquist (1988) suggested a method which is much more effective. If one is taking only monopoles into account, then only the centre of mass is of interest; it is not important how the charge

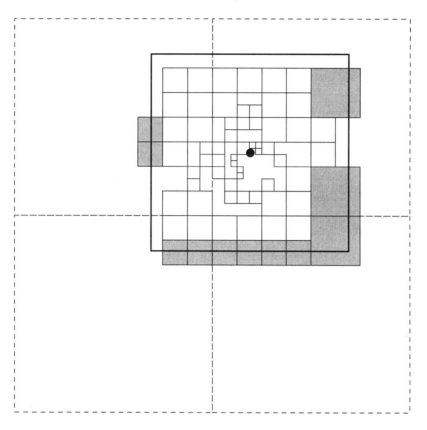

Fig. 5.3. Straddling of pseudoparticles across the minimum image boundary.

is distributed within the cell of the pseudoparticle. Assuming the mass of the pseudoparticle is homogeneously distributed in a cube, Bouchet and Hernquist take into account the overlap with another cube centred around the particle (see Fig. 5.4). This is analogous to the 'Cloud-in-Cell' method used in PIC codes. Due to the even mass distribution, the new centres of mass of these overlaps can then be used in the force calculation.

This method gives rise to two additional sources of error:

- The error due to the exclusion of terms outside the minimum image
- The splitting of the cells straddling boundaries while retaining only monopole terms

The differences between the minimum image method and the Ewald summation due to the first effect arise mainly at small wavenumbers, that is, large

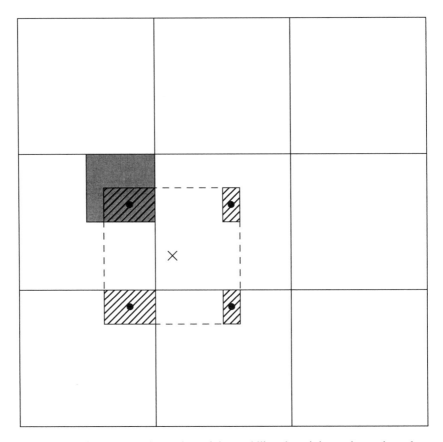

Fig. 5.4. Folding a 'corner' pseudoparticle straddling the minimum image boundary.

scale lengths, whereas the behaviour on small scales is nearly identical. (For a more detailed discussion, see Bouchet and Hernquist [1988].) Difficulties due to the minimum image approximation will therefore mainly arise for strongly coupled systems.

However, the more severe problem with this method is that higher order terms cannot be included in the multipole expansion straightforwardly. Therefore, problems that need a high accuracy cannot be handled by this method. This not only means that the path of a single particle is not correct, but due to the periodic nature, drift instabilities can occur, which eventually lead to a breakdown of the simulation.

One way out of this would be to borrow an idea from the FFM algorithm (see Chapter 7) and perform a local expansion of the 'straddle-cells'. This would be possible because one knows not only the monopole but the dipole and quadrupole moments of these cells too.

5.2 Ewald Summation

In view of the added complexity in applying the minimum image method to tree codes, and because in general one would like to be able to tackle problems with arbitrary Γ anyway, implementing the full Ewald sum instead seems to be worth the effort for most problems. The basic idea of the summation is to split the slowly convergent series of Eq. 5.1 into two rapidly convergent ones in real and reciprocal space. This method was originally introduced by Ewald (Ewald 1921) to determine the lattice constants of regular crystals. The idea was first taken up for use in Monte Carlo simulation codes by Brush, Sahlin, and Teller (1966) and detailed descriptions of the method can be found in Sangster and Dixon (1976) and Hansen (1985). The Ewald summation has been previously implemented in tree codes by Hernquist, Bouchet, and Suto (1991). In FMM codes, this extension is made by shifting the n-term multipole expansion for the whole system to its periodic images (Schmidt & Lee 1991) – see Chapter 7.

For our tree code we have chosen a slightly different notation in order to generalise the method for the multipole expansion. If we just look at a single particle p, its potential due to the other particles at positions $\mathbf{r}_i$ and their periodic images at $\mathbf{r_n} - \mathbf{r}_i$ is

$$\Phi_p = \sum_{\mathbf{n}} \sum_i \frac{q_i}{|\mathbf{r_n} - \mathbf{r}_i|},$$

where $\mathbf{n}$ is an integer vector denoting the images of the root cell.

Applying the standard Ewald procedure, this is replaced by the two rapidly converging series

$$\Phi_p = \Phi_I + \Phi_{II}$$

$$= \sum_{\mathbf{n}} \sum_i q_i \frac{\text{erfc}(\alpha r_{\mathbf{n}i})}{r_{\mathbf{n}i}} \tag{5.2}$$

$$+ \frac{1}{\pi L} \sum_i \sum_{\mathbf{h} \neq 0} q_i \exp\left(\frac{-\pi^2 |h|^2}{\alpha^2 L^2}\right) \cos\left(\frac{2\pi}{L} \mathbf{h} \cdot \mathbf{r}_{oi}\right),$$

where erfc is the complementary error function and $\mathbf{h}$ is a reciprocal lattice vector in units such that its components are integers. We have also defined $\mathbf{r}_{\mathbf{n}i} = \mathbf{r_n} - \mathbf{r}_i$; $\mathbf{r}_{oi} = \mathbf{r}_o - \mathbf{r}_i$. The parameter α is arbitrary, and although Φ_I and Φ_{II} depend on the choice of α, the sum Φ_p is independent of it.

However, α determines the relative convergence rate of the two series. In the actual numerical calculation the infinite sums over $\mathbf{n}$ and $\mathbf{h}$ are truncated, and the constant $\alpha \simeq 2/L$ is a parameter which optimises the convergence rates of the two series.

So in standard particle–particle codes, simply by substituting the $1/r$-potential by the Ewald sum given by Eq. 5.2 one can include the effect of an infinite array of images of particles. The equivalent expression for the force is

$$F_x^p = F_x^I + F_x^{II}$$

$$= \sum_{\mathbf{n}} \sum_i \frac{q_i \mathbf{x}_{\mathbf{n}i}}{r_{\mathbf{n}i}^3} \left[\operatorname{erfc}(\alpha r_{\mathbf{n}i}) + \frac{2\alpha r_{\mathbf{n}i}}{\sqrt{\pi}} \exp(-\alpha^2 r_{\mathbf{n}i}^2) \right]$$

$$+ \frac{2}{L^2} \sum_i \sum_{\mathbf{h} \neq 0} q_i h_x \exp\left(\frac{-\pi^2 h^2}{\alpha^2 L^2} \right) \sin\left(\frac{2\pi}{L} \mathbf{h} \cdot \mathbf{r}_{os} \right). \qquad (5.3)$$

Although the sums in Eqs. 5.2 and 5.3 can be calculated with high accuracy their evaluation is relatively time consuming. There are several ways to reduce the computation time of these sums (Sangster & Dixon 1976), of which the most common is to tabulate the correction of the Coulombic force due to the periodic images. This is done by constructing a three-dimensional grid of the size of the simulation box. Calculating the force between a particle at the centre of this cube and an imaginary particle at a grid point including all its images, using (5.3), the force correction F_{corr} is tabulated by subtracting the Coulombic force F_{Coul} from the Ewald force F_{Ewald}. Figure 5.5 shows the correction field for a particle sitting at the bottom left corner of the cell.

This tabulation has to be done only once before the actual simulation starts. Thereafter, it can be used during the simulation to evaluate the force corrections at the actual positions of the particles by a simple linear interpolation. In this way, the periodic images are included by adding a correction term to the Coulombic force,

$$F(x) = F_{Coul}(x) + F_{corr}(x).$$

As Fig. 5.5 shows, the correction field actually varies quite slowly over the relevant simulation region, so we use a relatively low order interpolation scheme. This somewhat cumbersome tabulation-interpolation procedure does indeed save much computing time over the exact evaluation of the Ewald sums.

When implementing the Ewald method into the tree code, we deal with particle–pseudoparticle interactions rather than particle–particle interactions. In Eqs. 5.2 and 5.3 this can be taken into account by q_i representing the sum of the charges of the pseudoparticle and $\mathbf{r}_i$ representing the centre of charges.

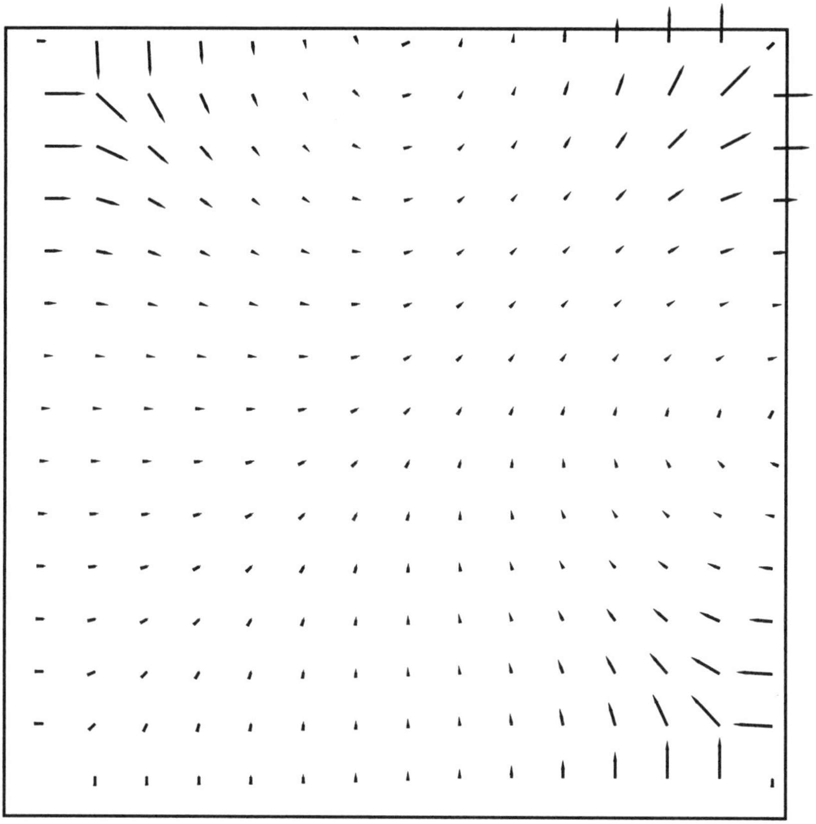

Fig. 5.5. Ewald correction field for a particle at $\mathbf{r} = (0, 0, 0)$.

Choosing the notation according to Fig. 5.6 the centres of charges are given by

$$\mathbf{r}_{coc} = \frac{\sum_i |q_i| \mathbf{r}_i}{\sum_i |q_i|}.$$

The Ewald method has been implemented in tree codes (Bouchet & Hernquist 1988) and in FMM codes (Schmidt & Lee 1991) for cosmological applications, but for many applications it is not sufficient to include just the monopole moment of the pseudoparticle in the force calculation. The accuracy of the calculation can be improved significantly by including higher moments of the multipole expansion as follows (see Pfalzner and Gibbon 1994).

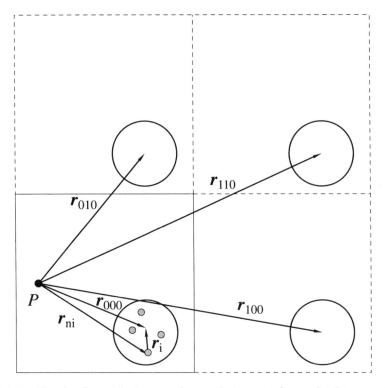

Fig. 5.6. Notation for multipole expansion: $\mathbf{r}_{\mathbf{n}i}$ is the vector from point P to a specific particle in the charge cluster, $\mathbf{r}_{\mathbf{n}}$ is the vector from P to the centre of charge of the cluster, and $\mathbf{r}_i$ is the vector from the particle to the centre of charge.

Expanding the potential Φ_p about the centre of charge

$$
\Phi_p = \sum_{\mathbf{n}} \sum_i q_i \left(1 - x_i \frac{\partial}{\partial x} - y_i \frac{\partial}{\partial y} - z_i \frac{\partial}{\partial z} \right.
$$
$$
+ \frac{x_i^2}{2} \frac{\partial^2}{\partial x^2} + \frac{y_i^2}{2} \frac{\partial^2}{\partial y^2} + \frac{z_i^2}{2} \frac{\partial^2}{\partial z^2}
$$
$$
\left. + x_i y_i \frac{\partial^2}{\partial x \partial y} + y_i z_i \frac{\partial^2}{\partial y \partial z} + z_i x_i \frac{\partial^2}{\partial z \partial x} + \ldots \right) \Phi(\mathbf{r_n}),
$$

we obtain:

$$
\Phi_p = M a_0 + D_x a_x + D_y a_y + D_z a_z
$$
$$
+ \frac{1}{2} Q_{xx} a_{xx} + \frac{1}{2} Q_{yy} a_{yy} + \frac{1}{2} Q_{zz} a_{zz}
$$
$$
+ Q_{xy} a_{xy} + Q_{yz} a_{yz} + Q_{zx} a_{zx}, \tag{5.4}
$$

where:

$$a_0 = \sum_{\mathbf{n}} \frac{\text{erfc}(\alpha r_{\mathbf{n}})}{r_{\mathbf{n}}} + \frac{1}{\pi L} \sum_{\mathbf{h} \neq 0} A(h) \cos\left(\frac{2\pi}{L}\mathbf{h} \cdot \mathbf{r_0}\right),$$

$$a_x = \sum_{\mathbf{n}} \frac{x_{\mathbf{n}} B_1}{r_{\mathbf{n}}^3} + \frac{2}{L^2} \sum_{\mathbf{h} \neq 0} h_x A(h) \sin\left(\frac{2\pi}{L}\mathbf{h} \cdot \mathbf{r_0}\right),$$

$$a_y = \sum_{\mathbf{n}} \frac{y_{\mathbf{n}} B_1}{r_{\mathbf{n}}^3} + \frac{2}{L^2} \sum_{\mathbf{h} \neq 0} h_y A(h) \sin\left(\frac{2\pi}{L}\mathbf{h} \cdot \mathbf{r_0}\right),$$

$$a_z = \sum_{\mathbf{n}} \frac{z_{\mathbf{n}} B_1}{r_{\mathbf{n}}^3} + \frac{2}{L^2} \sum_{\mathbf{h} \neq 0} h_z A(h) \sin\left(\frac{2\pi}{L}\mathbf{h} \cdot \mathbf{r_0}\right),$$

$$a_{xx} = \sum_{\mathbf{n}} \left(\left(\frac{3x_{\mathbf{n}}^2}{r_{\mathbf{n}}^5} - \frac{1}{r_{\mathbf{n}}^3}\right) B_1 + \frac{4x_{\mathbf{n}}^2}{r_{\mathbf{n}}^2} B_2 \right)$$
$$- \frac{4\pi}{L^3} \sum_{\mathbf{h} \neq 0} h_x^2 A(h) \cos\left(\frac{2\pi}{L}\mathbf{h} \cdot \mathbf{r_0}\right),$$

$$a_{yy} = \sum_{\mathbf{n}} \left(\left(\frac{3y_{\mathbf{n}}^2}{r_{\mathbf{n}}^5} - \frac{1}{r_{\mathbf{n}}^3}\right) B_1 + \frac{4y_{\mathbf{n}}^2}{r_{\mathbf{n}}^2} B_2 \right)$$
$$- \frac{4\pi}{L^3} \sum_{\mathbf{h} \neq 0} h_y^2 A(h) \cos\left(\frac{2\pi}{L}\mathbf{h} \cdot \mathbf{r_0}\right),$$

$$a_{zz} = \sum_{\mathbf{n}} \left(\left(\frac{3z_{\mathbf{n}}^2}{r_{\mathbf{n}}^5} - \frac{1}{r_{\mathbf{n}}^3}\right) B_1 + \frac{4z_{\mathbf{n}}^2}{r_{\mathbf{n}}^2} B_2 \right)$$
$$- \frac{4\pi}{L^3} \sum_{\mathbf{h} \neq 0} h_z^2 A(h) \cos\left(\frac{2\pi}{L}\mathbf{h} \cdot \mathbf{r_0}\right),$$

$$a_{xy} = \sum_{\mathbf{n}} \left(\frac{3x_{\mathbf{n}} y_{\mathbf{n}}}{r_{\mathbf{n}}^5} B_1 + \frac{4x_{\mathbf{n}} y_{\mathbf{n}}}{r_{\mathbf{n}}^2} B_2 \right)$$
$$- \frac{4\pi}{L^3} \sum_{\mathbf{h} \neq 0} h_x h_y A(h) \cos\left(\frac{2\pi}{L}\mathbf{h} \cdot \mathbf{r_0}\right),$$

$$a_{yz} = \sum_{\mathbf{n}} \left(\frac{3y_{\mathbf{n}} z_{\mathbf{n}}}{r_{\mathbf{n}}^5} B_1 + \frac{4y_{\mathbf{n}} z_{\mathbf{n}}}{r_{\mathbf{n}}^2} B_2 \right)$$
$$- \frac{4\pi}{L^3} \sum_{\mathbf{h} \neq 0} h_y h_z A(h) \cos\left(\frac{2\pi}{L}\mathbf{h} \cdot \mathbf{r_0}\right),$$

$$a_{zx} = \sum_{\mathbf{n}} \left(\frac{3z_{\mathbf{n}} x_{\mathbf{n}}}{r_{\mathbf{n}}^5} B_1 + \frac{4z_{\mathbf{n}} x_{\mathbf{n}}}{r_{\mathbf{n}}^2} B_2 \right)$$
$$- \frac{4\pi}{L^3} \sum_{\mathbf{h} \neq 0} h_z h_x A(h) \cos \left(\frac{2\pi}{L} \mathbf{h} \cdot \mathbf{r_0} \right),$$

$$A(h) = \exp \left(\frac{-\pi^2 |h|}{\alpha^2 L^2} \right),$$

$$B_1(r_{\mathbf{n}}) = \operatorname{erfc}(\alpha r_{\mathbf{n}}) + \frac{2\alpha r_{\mathbf{n}}}{\pi^{\frac{1}{2}}} \exp(-\alpha^2 r_{\mathbf{n}}^2),$$

$$B_2(r_{\mathbf{n}}) = \frac{\alpha^3}{\pi^{\frac{1}{2}}} \exp(-\alpha^2 r_{\mathbf{n}}^2). \tag{5.5}$$

The multipole moments are defined as

$$M = \sum_i q_i; \quad \mathbf{D} = \sum_i q_i \mathbf{r}_i; \quad Q_{xx} = \sum_i q_i x_i^2; \quad Q_{xy} = \sum_i q_i x_i y_i,$$

and so forth. The Ewald force field can be found in an analogous manner. The x-component of

$$\mathbf{F}_p(r) = \sum_{\mathbf{n}} \frac{q_p q_s \mathbf{r}_{\mathbf{n}i}}{r_{\mathbf{n}i}^3} \tag{5.6}$$

becomes

$$F_x^p / q_p = M a_x + D_x a_{xx} + D_y a_{xy} + D_z a_{xz}$$
$$+ \frac{1}{2} Q_{xx} a_{xxx} + \frac{1}{2} Q_{yy} a_{yyx} + \frac{1}{2} Q_{zz} a_{zzx}$$
$$+ Q_{xy} a_{xxy} + Q_{yz} a_{xyz} + Q_{zx} a_{xxz}, \tag{5.7}$$

where a_x, a_{xx}, a_{xy}, etc., and the multipole moments are defined as before, and

$$a_{xxx} = \sum_{\mathbf{n}} \left(\left(\frac{15 x_{\mathbf{n}}^3}{r_{\mathbf{n}}^7} - \frac{9 x_{\mathbf{n}}}{r_{\mathbf{n}}^5} \right) B_1 + \left(\frac{20 x_{\mathbf{n}}^3}{r_{\mathbf{n}}^4} - \frac{12 x_{\mathbf{n}}}{r_{\mathbf{n}}^2} + \frac{8\alpha^2 x_{\mathbf{n}}^3}{r_{\mathbf{n}}^2} \right) B_2 \right)$$
$$- \frac{8\pi^2}{L^4} \sum_{\mathbf{h} \neq 0} h_x^3 A(h) \sin \left(\frac{2\pi}{L} \mathbf{h} \cdot \mathbf{r_0} \right),$$

$$a_{xxy} = \sum_{\mathbf{n}} \left(\left(\frac{15 x_{\mathbf{n}}^2 y_{\mathbf{n}}}{r_{\mathbf{n}}^7} - \frac{3 y_{\mathbf{n}}}{r_{\mathbf{n}}^5} \right) B_1 + \left(\frac{20 x_{\mathbf{n}}^2 y_{\mathbf{n}}}{r_{\mathbf{n}}^4} - \frac{4 y_{\mathbf{n}}}{r_{\mathbf{n}}^2} + \frac{8\alpha^2 x_{\mathbf{n}}^2 y_{\mathbf{n}}}{r_{\mathbf{n}}^2} \right) B_2 \right)$$
$$- \frac{8\pi^2}{L^4} \sum_{\mathbf{h} \neq 0} h_x^2 h_y A(h) \sin \left(\frac{2\pi}{L} \mathbf{h} \cdot \mathbf{r_0} \right),$$

$$a_{xyz} = \sum_{\mathbf{n}} \left(\frac{15 x_{\mathbf{n}} y_{\mathbf{n}} z_{\mathbf{n}}}{r_{\mathbf{n}}^7} B_1 + \left(\frac{20 x_{\mathbf{n}} y_{\mathbf{n}} z_{\mathbf{n}}}{r_{\mathbf{n}}^4} + \frac{8\alpha^2 x_{\mathbf{n}} y_{\mathbf{n}} z_{\mathbf{n}}}{r_{\mathbf{n}}^2} \right) B_2 \right)$$

$$- \frac{8\pi^2}{L^4} \sum_{\mathbf{h} \neq 0} h_x h_y h_z A(h) \sin \left(\frac{2\pi}{L} \mathbf{h} \cdot \mathbf{r_0} \right).$$

There are ten of these quantities in all, the remainder of which can be found by cyclic rotation. The calculation of the dipole and quadrupole moments requires a little care, because they are defined relative to the centre of charge. Nevertheless, as with the open-boundary expansion described in Chapter 2 (see Fig. 2.11), we can still make use of the moments of the daughter cell to calculate the parent's moments. We recall that each individual $\mathbf{r}_i$ in the sum of a daughter moment is shifted by the same vector $\mathbf{r}_{sd}$. This means that

instead of: we require:

$\sum_i q_i x_i$ $\sum_i q_i x_i - x_{sd} \sum_i q_i$

$\sum_i q_i x_i^2$ $\sum_i q_i x_i^2 - x_{sd} \sum_i q_i x_i + x_{sd}^2 \sum_i q_i$

$\sum_i q_i x_i y_i$ $\sum_i q_i x_i y_i - x_{sd} \sum_i q_i y_i - y_{sd} \sum_i q_i x_i - x_{sd} y_{sd} \sum_i q_i$

and so forth. The sums $\sum_i q_i$, $\sum_i q_i x_i$, $\sum_i q_i x_i^2$, and $\sum_i q_i x_i y_i$ and the equivalent ones for the other spatial directions are evaluated together with the centres of charge, including these shifts, starting at the leaves and working down to the root. These results are used later to calculate the contribution in force.

Although these sums converge rapidly, again it takes considerably more computational effort than the $1/r$-multipole expansion with open boundaries; therefore it is useful to tabulate the higher moment *corrections* to the potential and fields within the simulation volume too. Subtracting the multipole terms obtained from expanding the Coulomb potential,

$$T_0 = a_0 - \frac{1}{r},$$

$$T_x = a_x - \frac{x}{r^3},$$

$$T_y = a_y - \frac{y}{r^3},$$

$$T_z = a_z - \frac{z}{r^3},$$

$$T_{xx} = a_{xx} - \left(\frac{3x^2}{r^5} - \frac{1}{r^3} \right),$$

$$T_{yy} = a_{yy} - \left(\frac{3y^2}{r^5} - \frac{1}{r^3}\right),$$

$$T_{zz} = a_{zz} - \left(\frac{3z^2}{r^5} - \frac{1}{r^3}\right),$$

$$T_{xy} = a_{xy} - \frac{3xy}{r^5},$$

$$T_{yz} = a_{yz} - \frac{3yz}{r^5},$$

$$T_{zx} = a_{zx} - \frac{3zx}{r^5},$$

$$T_{xxx} = a_{xxx} - \left(\frac{15x^3}{r^7} - \frac{9x}{r^5}\right),$$

$$T_{yyy} = a_{yyy} - \left(\frac{15y^3}{r^7} - \frac{9y}{r^5}\right),$$

$$T_{zzz} = a_{zzz} - \left(\frac{15z^3}{r^7} - \frac{9z}{r^5}\right),$$

$$T_{xxy} = a_{xxy} - \left(\frac{15x^2y}{r^7} - \frac{3y}{r^5}\right),$$

$$T_{yyx} = a_{yyx} - \left(\frac{15y^2x}{r^7} - \frac{3x}{r^5}\right),$$

$$T_{xxz} = a_{xxz} - \left(\frac{15x^2z}{r^7} - \frac{3z}{r^5}\right),$$

$$T_{zzx} = a_{zzx} - \left(\frac{15z^2x}{r^7} - \frac{3x}{r^5}\right),$$

$$T_{yyz} = a_{yyz} - \left(\frac{15y^2z}{r^7} - \frac{3z}{r^5}\right),$$

$$T_{zzy} = a_{zzy} - \left(\frac{15z^2y}{r^7} - \frac{3y}{r^5}\right),$$

$$T_{xyz} = a_{xyz} - \frac{15xyz}{r^7}. \tag{5.8}$$

These coefficients are multiplied by the appropriate multipole moments to give a force or potential, as required. All twenty of these quantities are needed to tabulate the correction fields to quadrupole precision (but only the first ten are needed for the correction to dipole precision). The actual force is then found by summing over the particles within the physical simulation volume and then adding a correction F_{corr}, obtained by interpolating from the eight nearest tabulated values. When the tree algorithm is used together with the

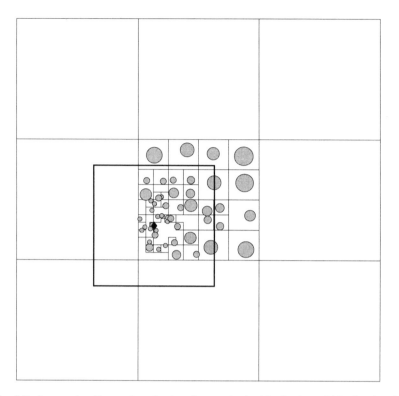

Fig. 5.7. Interaction list produced using the standard s/d criterion within the simulation region.

Ewald correction table, the interaction list is modified by gathering terms within a volume $\pm L/2$ centred on the particle.

The standard interaction list used for open-boundary problems looks something like Fig. 5.7: All terms sit within the 'primary' cell or simulation region. If we now 'wrap' the terms sitting farther than $L/2$ from the particle, the resulting list is lopsided, and may contain some very large pseudoparticles close to the particle. In fact, only particles at the centre of the simulation box will have sensible interaction lists. To avoid this, we must instead use the criterion:

$$s/d_{min} < \theta,$$

where

$$d_{min} = (x_{min}^2 + y_{min}^2 + z_{min}^2)^{\frac{1}{2}},$$

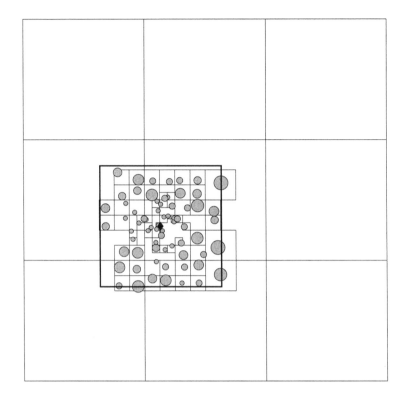

Fig. 5.8. Interaction list modified to take proper account of periodic boundaries.

$$x_{min} = \min(x, x - L, x + L)$$
$$y_{min} = \min(y, y - L, y + L)$$
$$z_{min} = \min(z, z - L, z + L).$$

Figure 5.8 shows the resulting modified list. This ensures that the corrections remain small for the particle's nearest (periodic) neighbours, which would not be the case if the usual 'open-boundary' interaction list were taken. In practice, we find that for $\theta < 0.75$, it is often sufficient to take a quadrupole expansion for the terms inside the minimum image cell, with corrections evaluated to *dipole* accuracy.

5.3 Timing

In Table 5.1 we have compared our BH tree algorithm with a PP version (both vectorised) for a randomly initialised electron-ion plasma. The timings were

Table 5.1. *Comparison between Barnes-Hut-Ewald and PP codes. The 'exact' PP timings are taken from Schmidt and Lee (1991)*

	PP		BHE	
N	Ewald table	Ewald sum	$\theta = 0.5$	$\theta = 0.75$
250	0.039	.	.	0.10
500	0.149	.	0.666	0.25
1,000	0.588	8.61	1.481	0.60
2,000	2.265	29.62	3.634	1.42
4,000	8.994	109.3	9.115	3.29
10,000	55.625	624.4	29.18	9.96
20,000	222.5	2552.	62.74	21.4
40,000	890.0	8101.	133.69	46.9

made on an IBM3090-60J, but renormalised to Cray–YMP time to facilitate comparison with the 3D FMM code of Schmidt and Lee (1991). The accuracy of the force calculation was estimated according to

$$\varepsilon_k = \left\{ \frac{\sum_{i=1}^{N_{test}} \left(F_{ki}^{tree} - F_{ki}^{direct} \right)^2}{\sum_{i=1}^{N_{test}} \left(F_{ki}^{tree} \right)^2} \right\}^{\frac{1}{2}},$$

where F_{ki}^{tree} and F_{ki}^{direct} are the kth components of the forces on particle i, evaluated using the tree and direct methods, respectively. The average over the three force components x, y, and z is given in the bottom row of the table. The force and potential were evaluated to quadrupole precision (using terms gathered from the minimum image cell – see Fig. 5.2) for two different θ and corrected to dipole precision with the Ewald sums tabulated on a 15^3 grid. This was sufficient to achieve the same degree of energy conservation as PP (using the same correction grid) for a plasma with coupling parameter $\Gamma < 1$. Performing the Ewald force correction to quadrupole precision roughly doubles the time shown in Table 5.1.

The force errors for the periodic code are determined largely by the resolution of the correction grid, which gives about 1% accuracy in this case. This can be reduced to that of the open boundary code by using a finer grid, but for a dynamic application there is little point in computing forces to a higher accuracy if the integration scheme conserves energy to only 1 or 2%. We have therefore measured the code against an optimised PP version which also makes use of a tabulated Ewald correction, and is thus much faster than the 'exact' PP summation timings quoted from Schmidt and Lee (1991) in Table 5.1.

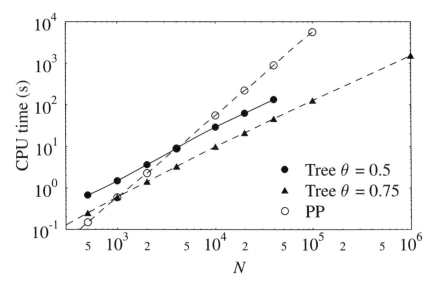

Fig. 5.9. Performance of the $N \log N$ algorithm compared with direct summation (PP).

In order to get a better idea of how the two methods compare, we have plotted in Fig. 5.9 the PP and tree results from Table 5.1. The periodic tree code is 3–4 times slower than the open-boundary version. For plasma simulations one typically takes $N = 10^4 - 10^5$ to obtain reasonable statistics in the velocity distribution and density; from Fig. 5.9 we see that the periodic tree code is between 10 and 100 times quicker than PP.

5.4 Monte Carlo Application

In the previous chapters we have seen how the dynamics of a system can be described using a tree algorithm. This dynamical behaviour is then used to investigate the inherent – equilibrium or nonequilibrium – properties of this system. These molecular dynamics (MD) simulations obtain the macroscopic quantities by averaging over time. If one is interested only in the equilibrium properties, a different method is used: the Monte Carlo (MC) method (Metropolis et al. 1953, Sangster & Dixon 1976, Hansen & McDonald 1981). In contrast to molecular dynamics, the MC method calculates the features of the system by averaging over ensembles. However, in MC calculations the moves of the particles are artificial rather than dynamical so that only the average positions of the particles are meaningful. It is for this reason that only equilibrium properties can be evaluated. If one is interested only in such static variables as the equation of state and pair correlation functions, the MC method has some advantages over molecular dynamics methods:

- The temperature is a fixed quantity, whereas in the MD method it is the total energy which is conserved and the temperature fluctuates.
- There are no problems of close collisions and reduction in the timestep.
- An extension to include many-body terms in the potential would be easier by the MC technique.

However, the MC method has the disadvantage that it cannot deal with the dynamical features of a system. In this book we have been mainly interested in new problems which tree codes allow us to tackle. One such area where there are many open questions is the dynamics and transport of nonequilibrium systems, and therefore we will describe only briefly how tree codes could be used for MC calculations.

The MC method was first proposed by Metropolis et al. (Metropolis et al. 1953) to evaluate the properties of interacting individual molecules and it was further developed by Wood (1968). Since then the MC method has found widespread applications in all fields of physics; for an overview of recent applications, see, for example, Binder (1987). The tree code could be used to speed up all MC calculations that include a long-range potential.

In the original version of the MC method, only two-body interactions were taken into account and normalized canonical ensemble averages were computed. Since then the method has been extended to cases of isobaric and grand canonical ensembles. Here we use the original version to implement with the tree algorithm, but the procedure is exactly the same for isobaric and grand canonical ensembles, just replacing $\exp(-\Delta E_{pot}/kT)$ by the appropriate equivalent expressions.

The potential energy of the system is given by

$$E_{pot} = \frac{1}{2} \sum_{i=1}^{N} \sum_{j=1}^{N} V_{ij}(r_{ij}). \tag{5.9}$$

The potential energy of the system can be calculated similarly to the force calculation using the tree structure. As in the force calculation the criterion $s/d < \theta$ decides whether the potential energy between a specific particle and a pseudoparticle is calculated or the pseudoparticle is resolved into its daughter cells. The potential energy is calculated by taking into account the immediate neighbourhood directly and coarser groupings of particles at larger distances.

MC methods can be used only to investigate equilibrium properties F of a system, as given by

$$\langle F \rangle = \frac{\int F \exp(-E_{pot}/kT)d^3\mathbf{r}d^3\mathbf{p}}{\int \exp(-E_{pot}/kT)d^3\mathbf{r}d^3\mathbf{p}}. \tag{5.10}$$

The Monte Carlo method integrates over a random sampling of points. If one would choose just a random configuration, it would be very likely that the factor $\exp(-E_{pot}/kT)$ is very low and would therefore have a low weight. In order to avoid this, one chooses configurations with probability $\exp(-E_{pot}/kT)$ and weights them evenly. This is done in the following way.

Each particle is moved according to $\mathbf{r}_i \rightarrow \mathbf{r}_i + \psi\alpha$, where α is the maximum allowed displacement and each component of ψ is a random number between -1 and 1. If the change of energy ΔE_{pot} that such a move would cause is positive, this move will not be performed. If, on the other hand, $\Delta E_{pot} < 0$, a random number ϕ is chosen and only if $\phi < \exp(-\Delta E_{pot}/kT)$ is the move eventually allowed. This method allows configurations to be chosen with a probability $\exp(-E_{pot}/kT)$.

The basic idea is to choose the system with minimal potential energy of the system by random movement of the particles. The tree method for this procedure can be summarised as follows.

> Construct tree
> Evaluate potential energy
> 1. Move one particle
> 2. Construct tree
>> Evaluate energy
>> Compare old and new energy
>> **if** E(new) < E(old) or ψ < exp(-Δ E/kT)
>>> accept move
>> **else**
>>> move particle back to old position
>> **endif**
> goto 1

As in the force calculation we have an approximate $N \log N$ dependence of the computation time per move.

5.5 Nonequilibrium Systems

So far we have considered only the simulation of systems in equilibrium. The external conditions had no influence on any thermodynamical fluxes. Usually transport coefficients are calculated using either linear response theory (McQuarrie 1976) or time-correlation functions (Allen & Tildesley 1987). This has two disadvantages: The statistical error is relatively large, and if the perturbation is large linear response theory is no longer applicable and time correlation functions have a poor signal-to-noise ratio. However, it is possible to investigate systems in a nonequilibrium state using MD methods and by measuring the response to a large perturbation directly.

The methods which are easiest to apply and interpret are those which are consistent with the usual periodic boundaries or modified in such a way that periodicity and translational invariance are preserved. For an overview see, for example, Allen and Tildsley (1987).

When a large perturbation is applied to a system it can respond either adiabatically, isothermally, or isoenergetically. If one considers an adiabatic system, applying a perturbation will result in heating the system. Although physically correct, this has the disadvantage that the state of the system changes as the simulation evolves. In order to sample over an isothermal system, one can either rescale the velocity distribution after a certain number of timesteps, or modify the usual equation of motion to take care of the heating. Both methods correspond physically to immersing the whole system in a heat bath.

In the case of velocity rescaling, assuming that the perturbation is oscillatory, this can be at every full cycle. Alternatively, the rescaling can be performed at every timestep. However, if one would use a tree code with multiple timescales, one should be careful to make sure the rescaling of all particle velocities is done at the same time. Various methods of doing this rescaling are described in Allen and Tildsley (1987).

If one chooses to include the external force in the equation of motion the constant kinetic temperature dynamics is generated by the following equations of motion (Ciccotti 1991):

$$\dot{r} = \frac{\mathbf{p}}{m} + \frac{\partial \mathbf{A}}{\partial p} \cdot \mathbf{F}_{ext}(t)$$

$$\dot{p} = \mathbf{f} - \frac{\partial \mathbf{A}}{\partial r} \cdot \mathbf{F}_{ext}(t),$$

where C and D describe the coupling of the system to the external force and a Hamiltonian perturbation was assumed. The Hamiltonian of the system contains an equilibrium term H_0 and a term which describes the perturbation H_p, thus

$$H = H_0 + H_p = H_0 + \mathbf{A}(\mathbf{r}, \mathbf{p}) \cdot \mathbf{F}_{ext}(t).$$

The compressibility condition for the system is then given by

$$\left[\frac{\partial}{\partial r} \frac{\partial \mathbf{A}}{\partial p} - \frac{\partial}{\partial p} \frac{\partial \mathbf{A}}{\partial r} \right] \cdot \mathbf{F}_{ext}(t) = 0.$$

Since most of these modifications involve velocity-dependent terms, it is usually easier to use the Gear algorithm than the Verlet algorithm – see Section 4.2. Alternatively, a leap-frog method with a modified velocity equation can be used:

$$\dot{\mathbf{r}}(t + \frac{1}{2}\delta t) = \dot{\mathbf{r}}(t - \frac{1}{2}\delta t) + (\mathbf{f}(t)/m - \psi \dot{\mathbf{r}}(t))\delta t.$$

6

Periodic Boundary Problems

In this chapter we will consider a variety of fields where the tree algorithm in combination with periodic boundaries can be applied and where the speedup in comparison to standard MD and MC codes enables previously inaccessible problems to be investigated. This set of applications is in no way exhaustive, but is intended to indicate the types of problems where the algorithm might best be put to use.

Practically every N-body MD or MC code could incorporate the tree algorithm. However, for systems with short-range potentials, like the Lennard–Jones potentials for the description of solids, this does not bring much of an advantage. Because the potential falls off very rapidly (see Fig. 6.1), a sharp cutoff can be used to exclude interactions of more distant particles whose contribution is negligible. The tree algorithm is mainly suited to systems with long-range forces such as Coulomb, where the summed effect of distant particles is important.

An ideal application is dense, fully-ionized plasmas. Here the particles are so closely packed that an analytical treatment is difficult, and many MD and MC calculations have been carried out to investigate their properties. Limitations in the number of simulation particles make some problems difficult to address due to a combination of poor statistics and small system size. The tree algorithm could be successfully applied here, because the particles interact purely through Coulomb forces.

Other applications are to be found in areas where the interaction potentials of the particles are more complex but contain long-range Coulombic contributions. Typical examples would be the simulation of ionic liquids, molten salts, macromolecules like protein, and metals with defects in materials science. Often the Coulombic part of the potential uses most of the computation time, therefore a better efficiency of this part of the calculation opens up new possibilities in this context too.

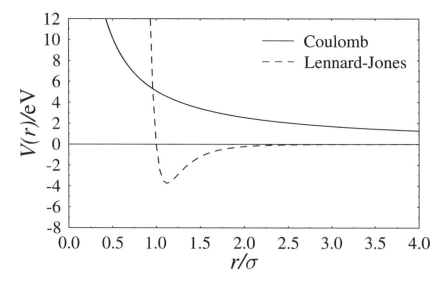

Fig. 6.1. Comparison of the Lennard–Jones and Coulomb potentials.

6.1 Plasma Physics: Collisions in Dense Plasmas

In contrast to the systems considered in Chapter 2, plasmas – because they are composed of oppositely charged particles – involve attractive as well as repulsive forces. Therefore, one requirement for many problems is charge neutrality, meaning $\sum_k (-1)^M q_k n_k = 0$, where M is the number of differently charged particles.

Plasmas span an extremely wide range of temperature and density as Fig. 6.2 shows. Depending on temperature and density, they exhibit different behaviour and require rather different theoretical modelling as a result.

An important parameter which characterises the strength of the interaction between the particles is the ion coupling parameter Γ. It essentially describes the ratio of the potential to the kinetic energy of the particles of the plasma

$$\Gamma = \frac{Z^2 e^2}{a k_b T}, \tag{6.1}$$

where Z is the charge of the particles, $a = (4\pi n_i/3)^{-1/3}$ is the ion sphere radius, and n_i is the ion density.

Most laboratory plasmas, like the ones in magnetic fusion experiments, have relatively low density but high temperatures, thus $\Gamma \ll 1$. The electrons and ions in these systems rarely collide and for most purposes can be treated like

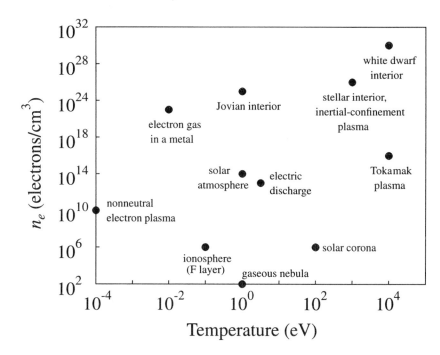

Fig. 6.2. Density temperature plot for various types of plasmas.

an ideal gas. The thermodynamic properties of such weakly coupled plasmas are analogous to those of an ideal gas.

One important difference of a system consisting of particles of opposite charges to single-charge systems is that a particle is shielded by the self-polarization cloud which it induces in the surrounding plasma. The range of this screening – the Debye length $\lambda_D = (k_B T / n_e e^2)^{1/2}$ – determines the collective behaviour of the system, resulting in a characteristic frequency ν_p called plasma frequency given by

$$\nu_p = \frac{\omega_p}{2\pi} = \left(\frac{2 n_e e^2}{m_e}\right)^{1/2}. \tag{6.2}$$

In comparison, the frequencies of the electron-electron and electron-ion collisions are given by

$$\nu_{ee} = 4\pi e^4 \frac{n_e}{m_e^2 v_e^3} \ln \Lambda_{ee}$$

$$\nu_{ei} = 4\pi Z^2 e^4 \frac{n_i}{m_e^2 v_e^3} \ln \Lambda_{ei}, \tag{6.3}$$

where $ln\,\Lambda_{\alpha\beta}$ is the Coulomb logarithm (Chen 1974) – are very small for plasmas with $\Gamma \ll 1$. In this case, the behaviour of the plasma is dominated by larger scale phenomena like instabilities and wave-particle interaction. These problems are best treated by mesh-based codes such as PIC, as discussed before. MD or MC simulations come into their own when large-angle collisional scattering cannot be neglected, meaning that $\Gamma \gtrsim 1$.

Such strongly coupled plasmas exist, for example, in planetary (Stevenson 1982) and stellar (Iyetomi & Ichimaru 1986) interiors, in white dwarfs (Chabrier et al. 1992), and in the crust of neutron stars. Experimentally they can be produced by high-intensity short-laser pulses (Perry & Mourou 1994) and in inertial confinement experiments (Lindl et al. 1992). Such plasmas have densities equivalent to that of solids and above, but temperatures in the range of 1 eV to 50 keV. Such plasmas cannot be regarded as an ideal gas anymore. The strong Coulomb coupling significantly affects the basic properties of the system, like the equation of state and transport coefficients. The reason for these changes is that at high densities, collisions become increasingly important.

Another important parameter to characterise the type of plasma is the degree to which the electron gas is degenerate. The electron degeneracy parameter is given by

$$\theta_e = \frac{k_B}{\epsilon_F}, \tag{6.4}$$

where $\epsilon_F = (\hbar^2(3\pi^2 n_e)^{2/3})/2m$ is the Fermi energy where n_e is the electron density. Figure 6.3 shows a temperature-density diagram for the ion coupling and the electron degeneracy parameter. For even higher densities the ions can be degenerate too, but we will not discuss this case here.

If the plasma has highly degenerate electrons, which is equivalent to the condition $T \ll T_F$, and the plasma density is so high that $a < (\pi/12)^{2/3}(\hbar/me^2)$, that is, the Fermi screening length is bigger than the mean inter-atomic spacing, then electrons can be treated as a uniform neutralizing background and only the ion dynamics need to be followed explicitly. This model is called the one-component plasma (OCP). For an overview of OCPs see (Ichimaru 1982). The potential energy of such systems in d dimensions is given by (Hansen 1985):

$$V_{pot}^{OCP} = \frac{1}{2|\Lambda|} \sum_{\mathbf{k}} \frac{C_d q^2}{k^2} \left[\sum_{i=1}^{N} \exp(i\mathbf{k} \cdot \mathbf{r}_i) \sum_{i=1}^{N} \exp(-i\mathbf{k} \cdot \mathbf{r}_i) - N \right],$$

where the summation over wave vectors compatible with the periodic boundary conditions indicates that the $k = 0$ component is cancelled by the background contributions. The first simulations of OCP plasmas were performed by Brush,

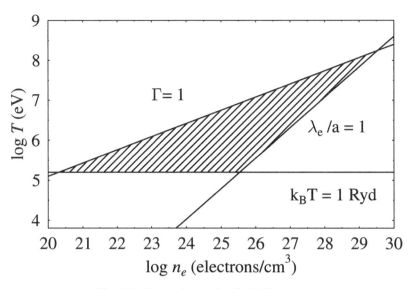

Fig. 6.3. Simulation region for OCP plasma.

Sahlin, and Teller (1966); later on, Hansen et al. (1979) and Slattery, Doolen, and DeWitt (1982) carried out similar calculations with improved accuracy. The OCP model can easily be extended to contain more than one ion species to study mixing effects, as, for example, in Baus (1977) and Hansen et al. (1979).

If, on the other hand, the electrons are only weakly degenerate, they have to be treated the same way as the ions in the simulation. Such plasmas fulfill simultaneously the conditions of strong coupling, weak degeneracy, and complete ionization. They are believed to exist, for example, at the initial stages of an imploding pellet in inertial confinement fusion. Treating attractive and repulsive forces at the same time in a two-component plasma (TCP) simulation requires some special care with a tree code. For single-charge systems (e.g., gravitational or OCP), the multipole expansion was used only in order to improve the accuracy of the calculation, whereas for two-component plasmas it is essential to include the higher moments of the multipole expansion (Pfalzner & Gibbon 1992). The reason is that the monopole moment is simply the sum of the charges in the pseudoparticle and it is quite likely that a pseudoparticle contains the same number of electrons as ions. If one were to perform the simulation only with monopole contributions, this pseudoparticle would be ignored completely.

Due to the attractive forces between the electrons and ions a 'bare' Coulomb potential would make the system unstable. Therefore, it is necessary to include a short-range truncation which allows this singularity to be dealt with. One usually uses effective pair potentials which can account for quantum diffraction as well as symmetry effects in an approximate way. Neglecting bound-state contributions and limiting scattering states to s-waves, which is valid for sufficiently high temperatures, the quantum-mechanical Slater sum reduces to an effective ion-ion and electron-ion pair potential of the form (Deutsch 1977)

$$V_{ij} = \frac{q_i q_j}{r} \left[1 - \exp\left(-\frac{r}{\lambda_{ij}}\right) \right], \tag{6.5}$$

with λ_{ij} being the deBroglie wavelength, which is given by

$$\lambda = \frac{\hbar}{\sqrt{2\pi m_{ij} k_B T}}, \tag{6.6}$$

where m_{ij} is the reduced mass. The effective electron-electron potential contains an additional term to take into account symmetry effects and is given by

$$V_{ee} = \frac{e^2}{r} \left[1 - \exp\left(-\frac{r}{\lambda_{ij}}\right) \right] + k_B T \ln 2 \exp\left(-\frac{r^2}{\pi \lambda_{ee}^2 \ln 2}\right). \tag{6.7}$$

The forces are, respectively, given by

$$\mathbf{F}_{ij} = \frac{q_i q_j}{r^3}\mathbf{r} - \frac{q_i q_j}{r^3}\mathbf{r}\left(\frac{1}{r} + \frac{1}{\lambda_{ij}}\right)\exp\left(-\frac{r}{\lambda_{ij}}\right)$$

$$\mathbf{F}_{ee} = \frac{e^2}{r^3}\mathbf{r} - \frac{q_i q_j}{r^3}\mathbf{r}\left(\frac{1}{r} + \frac{1}{\lambda_{ij}}\right)\exp\left(-\frac{r}{\lambda_{ij}}\right)$$
$$+ \frac{2k_B T}{\pi \lambda_{ee}}\mathbf{r}\exp\left(-\frac{r^2}{\pi \lambda_{ee}^2 \ln 2}\right).$$

Due to the large mass ratio m_i/m_e between the electrons and ions the dynamics of the two species occur on rather different timescales. The timestep has to be chosen according to the much quicker electron motion. Thus, little information about the ion dynamics is obtained.

The main property that has been investigated by MC and MD for dense plasmas is the radial distribution function or pair distribution function $g(r)$. This describes how likely it is to find a particle at a distance r from another particle, thus giving a measure for the spatial correlation between two particles, and is normalised so that $g(r)$ approaches unity as $r \to \infty$. In an ideal plasma the probability is independent of the distance, therefore $g(r) = 1$. In strongly coupled plasmas the repulsive forces between the ions and the attractive forces between the ions and electrons influence the radial distribution function. For

weakly coupled plasmas the radial distribution function can be obtained ana-
lytically:

$$g_{ii}(r) = 1 - \frac{e^2}{k_B T_i} \exp\left(-\sqrt{2}\left(\frac{4\pi n_i e^2}{k_B T_i}\right)^{1/2} r\right)$$

$$g_{ie}(r) = 1 + \frac{1}{4\pi n_e \lambda_{De}^2} \exp\left(-\frac{1}{\lambda_{De}^2} r\right). \tag{6.8}$$

The radial distribution is related to the static structure factor $S(k)$ via the Fourier
transform. $S(k)$ and $g(r)$ contain all the information necessary to calculate the
energy, pressure, and various thermodynamic functions of the system. They are
used to calculate the effects of external forces on the system, where basically
perturbation theory is applied. However, this is not possible if the external forces
are very strong and nonlinear effects become important. In this case, the effects
have to be measured directly in a system which contains the external force. In
order to obtain good statistics it is preferable to have as many particles in the
simulation box as possible. Therefore, tree methods would make new regions
accessible. To include external electric fields, like in the case of laser-plasma
interactions, is a straightforward procedure – it is simply an additional term in
the force calculation.

An external magnetic field $\mathbf{B}$ can be included in plasma simulation too. Bernu
(1981) studied the velocity autocorrelation function, and the self-diffusion of
the OCP in a strong, constant, external magnetic field. The equations of motion
involve the velocity-dependent Lorentz force and are given by

$$\mathbf{v}_i = \frac{d\mathbf{r}_i}{dt}$$

$$\mathbf{a}_i = \frac{q}{m}\mathbf{E}_i + \mathbf{v}_i \times \frac{q\mathbf{B}}{mc}. \tag{6.9}$$

The time reversible Lorentz force integrator is similar to the usual leap-frog
or Verlet algorithm:

$$\mathbf{r}_i^{(n+1)} - 2\mathbf{r}_i^{(n)} + \mathbf{r}_i^{(n-1)} = (\Delta t)^2 \left[\frac{q}{m}\mathbf{E}_i(r_i^n) + \mathbf{v}_i^{(n)} \times \frac{q\mathbf{B}}{mc}\right]$$

$$\mathbf{v}_i^{(n)} = \frac{1}{2\Delta t}\left[\mathbf{r}_i^{(n+1)} - \mathbf{r}_i^{(n-1)}\right].$$

Kwon et al. (1994) found that the standard OCP and tight-binding results
agree for $\Gamma \sim 2 - 6$, whereas they found differences for higher Γ.

As Fig. 6.4 shows, plasmas exist over a wide temperature-density range.
If the material is highly compressed but the temperatures are high too (of
the order of several eV) the classical OCP with bare ions in a homogeneous

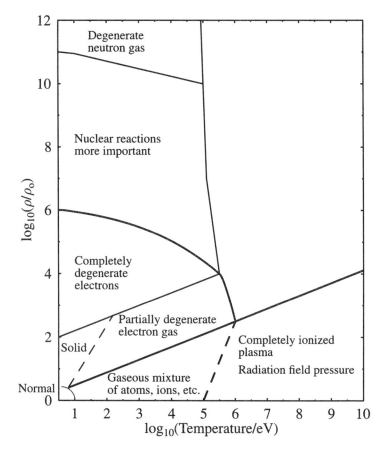

Fig. 6.4. Classification of different types of plasmas as a function of density and temperature.

background sea of electrons seems to be an adequate description. However, the transition region is difficult to access. Approaching it from the OCP side, quantal effects can be included by using Thomas–Fermi (Zérah et al. 1992) or effective potential (Dharma-wardana 1988) methods. One possibility to include electron-exchange, correlation, and temperature effects more precisely is the tight-binding method (Younger 1992).

Very strongly coupled plasmas ($\Gamma \sim 1$–20) can, for example, be found in white dwarfs (van Horn 1980) and the interior of Jupiter ($\Gamma \sim 20$–50) (Stevenson 1980) For extremely strong coupled plasmas, that is, in the region of high compression and low temperature, treating the electrons just as a homogeneous background is insufficient. More sophisticated techniques, including more elaborate potentials, have been developed, combining the classical treatment of the nuclei by Monte Carlo or molecular dynamics methods with the quantal treatment of the electrons based on Hartree–Fock, density-functional, and semi-empirical techniques (Ceperley & Alder 1987, Younger 1992, Hohl et al. 1993, Zérah et al. 1992). Plasmas with $1 < \Gamma < 180$ are called Coulomb fluids. The fluid freezes into a solid in the vicinity of $\Gamma = 150$–200.

However, a tree method in this case would not bring much advantage due to the fact that, in most cases, the quantal treatment requires a lot more time, so the number of particles in such simulations is not large enough to bring any advantage. But if the number of simulation particles reaches the region of several thousand an application would be favourable.

As Γ increases, the plasma reaches its crystallisation point and the short-range order becomes more pronounced so that the harmonic-lattice model starts to apply. One important question for very high Γ is where the fluid crystal transition or Wigner crystalisation takes place.

Metals can be regarded as electron-ion plasmas and OCP simulations have been carried out for liquid alkali metals. However, the difference between the timescale of electron and ion motions and the degeneracy of the conduction electrons requires more sophisticated MD simulation methods. (See, for example, Rahman [1964] and Turq [1977].)

More generally, the OCP model consists of a single species of charged particles embedded in a uniform background of neutralising charges. In this sense, the OCP model is used to simulate the system of valence electrons in a metal which is then regarded as a degenerate strongly coupled plasma.

6.2 Systems of More Complex Interaction Potentials

In the following, we discuss the application of the hierarchical tree method to systems which have a more complex interaction potential. In this case, the most effective way to reduce the computation time without compromising too much the accuracy of the simulation is the introduction of a cutoff parameter for the potential. It is obvious that a tree code is only advantageous if a long-range force, which in most of the examples will be a Coulombic force, is involved. The systems generally can be regarded neither as an ideal gas nor as a crystal structure. In addition to being in this theoretically difficult transition region, the systems consist of more than one species, so that chemical bonding forces play an important role.

Fluids are more strongly coupled, the forces are of a shorter-ranged kind, and therefore are usually modelled using Lennard–Jones potentials. As before, for this kind of system there is no real advantage using a hierarchical tree code.

However, for certain kinds of fluids long-range forces are essential, for example, in the simulation of charged ions ($V(r) \sim r^{-1}$) and dipolar molecules ($V(r) \sim r^{-3}$). In order to perform MD simulations for liquids one uses periodic boundary conditions unless very small liquid drops are investigated. The size of the system has to be chosen sufficiently large, because otherwise the periodic boundary conditions can introduce anisotropies in the fluid structure. For this kind of system the hierarchical tree method is a prime candidate to reduce the computation time. Two typical examples for liquid systems involving Coulombic forces are ionic crystals (Klein 1983) and molten salts (Hansen 1985).

Molten salts are quite similiar to two-component plasmas discussed in Section 6.1 in the sense that they consist of two species of opposite charges. However, in contrast to plasmas molten salts consist of oppositely charged ions and therefore the dynamical timescales of the two species are of the same order. In addition, they are usually much more strongly coupled ($\Gamma \sim 50$).

For alkali halides the simple potential (Adams 1975)

$$V_{ij} = \frac{e^2}{\sigma} \left[\frac{1}{b} \left(\frac{r_o}{r} \right)^n + Z_i Z_j \left(\frac{r_o}{r} \right) \right] \tag{6.10}$$

is used, where σ, b, and r_o are constants determined from experimental data. For rigid ions one usually takes a pair potential of the form (Tosi 1964)

$$V_{ij} = A_{ij} \exp(-r/n_{ij}) - C_{ij} \left(\frac{r_o}{r} \right)^6 - D_{ij} \left(\frac{r_o}{r} \right)^8 + \frac{Z_i Z_j e^2}{r}. \tag{6.11}$$

The constants $A_{i,j}$, $C_{i,j}$, $D_{i,j}$, and so forth are again determined from solid-state data. A detailed discussion of the potentials of molten and amorphous salts can be found in Vashishta et al. (1989, 1990). This potential consists of two-body and three-body interactions, where the two-body interaction includes a Coulombic term and takes account of steric repulsion due to the ionic size, charge dipole effects, and polarization. Polarized ions have been investigated by Sangster and Dixon (1976). The three-body interactions include bond-stretching and bond-bending terms. One typical material for such investigations is silica. This is of special interest because the Earth's crust is largely composed of silica and silicates (Jin et al. 1993, Zallen 1983).

Computer simulations that calculate the direct particle–particle interaction were first used to study chemical systems and molecular interactions, which actually resulted in calling the whole technique 'molecular dynamics', whichever kind of physical system is considered. Although molecules were the first physical systems to which this method was applied, there are still open questions in this field.

In this context, one major field of interest is polyatomic fluids (Beveridge et al. 1983, Jorgensen et al. 1981, Steinhauser 1981b) and watery solutions. It has been shown that hard-sphere (Alder & Wainwright 1959) or Lennard–Jones type (Rahman 1964) of potentials are insufficient to simulate many properties of such systems and the correct treatment of the long-range forces has emerged as one of the most important issues. The common technique to include Coulombic interactions to some extent and keep the simulation time down is to introduce a spherical cutoff, which in most simulations is on the order of 8Å. By varying this cutoff it has been shown that for neat water the energetic properties and the radial structure are not very sensitive to the correct treatment of the long-range forces, but the orientational structure and the dielectric constant are strongly affected (Stillinger & Rahman 1974, Pangali et al. 1980, Steinhauser 1981a). Only the employment of a reaction field or Ewald sum allows these quantities to be determined in a consistent way for neat polar systems (Belhadj et al. 1991). For systems of ions in water, employing cutoffs leads to significant deficiencies of the solution structure and thermodynamics (Brooks III. 1987). Interestingly, not only perturbations of the radial structure near the cutoff are observed but these perturbations propagate to the short-range structure of the water about the ions.

Another field of interest is lipid membrane systems and water in these systems, because the fluid, liquid crystal-like nature of the membrane makes experimental study of these systems difficult, especially at the atomic level (Murthy & Worthington 1991, Wiener et al. 1991, Alper et al. 1993). Typical particle numbers of simulations of such systems are several thousands to ten thousands and are therefore in the typical range where the use of a tree code can lead to a considerable speedup of the simulation. Zhou and Shoulten (1995) carried out an FMM calculation for a membrane-water interface with 32,800 atoms and included Coulomb forces between all atom pairs. They found that the polarization of the water is determined mainly by lipid head groups in the interfacial region.

In the case of polar fluids, Wertheim (1971) showed that most of the thermodynamic properties of a polar fluid can be expressed in terms of the sole projections g_{000}, g_{110}, and g_{112}. Interesting in the context of applying the tree method to this kind of system is the suggestion of Caillol (1992), who proposed using a simulation cell with the surface S_3 of a four-dimensional sphere (hypersphere) as an alternative to the standard Ewald method for periodic boundary conditions. The principal advantage is that the multipolar interactions have simple analytical expressions and one gets rid of the minimum image convention. Therefore, the simulation of a system of $\sim 1,000$ ions on a hypersphere is 3–4 times faster than the analogous simulation in the periodic boundary space. To what extent this technique could be combined with a hierarchical tree method and whether it could be used in other applications is a question well worth pursuing.

6.3 Biomolecules

Another application of molecular dynamics simulation where the utilisation of the hierarchical tree method could bring a huge advantage is in the modelling of biological macromolecules like protein. Proteins are of particular interest because they are one of the essential components of living systems, contributing to the structures of the organism and executing most of the tasks the organism needs to work properly.

MD simulations are an ideal tool to investigate biomolecules, because they have many things in common with the systems we discussed before: They contain many atoms, are densely packed, and work typically in a liquid environment. An overview of MD simulations of proteins can be found in Brooks et al. (1988), Clementi and Chakravorty (1990), and Levitt (1969); here we will review the basic ideas of MD simulations of proteins and the possibilities for applying the tree algorithm.

In order to perform a simulation the first question one has to address is: What does a protein look like? The size of proteins varies widely from approximately 1,000 to 10,000 atoms. Each protein consists of a polypeptide chain that is made of linked amino acids. This sequence of amino acids is called the primary structure and is the backbone of the protein. What distinguishes different proteins is the number of amino acids and its sequence in the polypeptide chain. The atoms that constitute the amino acids are basically H, C, N, O, and S. Approximately half of the atoms are hydrogen, but those are experimentally difficult to study, therefore most of the models focus on the positions of the heavier atoms. Experimental investigations show that the proteins have a secondary and a tertiary structure. The secondary structure is the typical folding of the protein, which shows certain regularities; the tertiary structure is the overall spatial arrangement of the amino acids in the protein.

For such complex systems it is not possible to obtain the quantum mechanical solution of the Schrödinger equation at the required speed to use in a dynamical simulation. Therefore, one usually chooses a reasonable potential with free parameters which are fitted to the experimental information available. The simplest and most widely used model for the interaction is the Coulomb plus

Lennard–Jones potentials or the 1-6-12-potential:

$$V_{nb} = \sum_{ij\,pairs} \frac{q_i q_j}{r} + 4\epsilon_{ij} \left[\left(\frac{\sigma_{ij}}{r}\right)^{12} - \left(\frac{\sigma_{ij}}{r}\right)^6 \right], \tag{6.12}$$

where ϵ_{ij} is the dispersion well depth and σ_{ij} is the Lennard–Jones diameter. So like in the systems before, the Coulombic part of the interaction can be calculated using the tree algorithm, reducing the computation time significantly, whereas the Lennard–Jones potential usually is a short-range force and can be calculated separately with a range cutoff. This potential is used to model the important nonbonded interactions between the atoms – the electrostatic interaction, the repulsive van der Waals term, and a part that models the dispersion attraction. In the calculation the intermolecular forces between molecules, the terms between all atom pairs are counted. Within the molecule, however, only forces between atoms separated by at least three bonds are included to take account of the fact that van der Waals spheres overlap considerably at the chemical bonding distance. The chemical bonding interactions are described by separate terms in the potential function.

In addition to the interactions described by Eq. 6.12 the bonding of the atoms plays an important role. For bound atoms, the so-called 1,2-pairs, a harmonic oscillator model is used, giving a potential of the form

$$V_{bond} = \sum_{1,2\text{-}pairs} \frac{1}{2} K_b (b - b_o)^2, \tag{6.13}$$

where K_b is the bond stretching force constant, b is the bond length, and b_o is the equilibrium distance parameter. The terms that keep the bond angles near the equilibrium geometry involve triples of atoms or neighbouring pairs of bonds. In its simplest form it is given by

$$V_{bond\ angle} = \sum_{bond\ angles} \frac{1}{2} K_\Theta (\Theta - \Theta_o)^2, \tag{6.14}$$

with K_Θ being the bending force constant, Θ the bond angle, and Θ_o the equilibrium value parameter. To model the hindered rotation about single and partial double bounds an explicit torsion potential is introduced:

$$V_{torsion} = \sum_{dihedral\ angles} K_\Phi \left[1 + \cos(n\Phi - \delta) \right], \tag{6.15}$$

where K_Φ is the dihedral angle force constant, Φ is the dihedral angle, n is multiplicity and δ is the phase.

At higher temperatures or pressures, not of general interest for biological macromolecules, rather more sophisticated expressions for the potential may be needed (Jackson 1986) and if velocity-dependent forces or temperature-scaling are incorporated, the calculation of the dynamics by the Verlet algorithm has to be modified to include the velocities in a similar manner to the inclusion of the magnetic field in the plasma physics section.

Not only does the potential used depend on experimental results; the initial setup of atoms in the polymer is obtained from X-ray measurements too. In contrast, the positions of the solvent atoms are obtained by fitting the biomolecules into a preequilibrated box of solvent atoms. The positions of all atoms are then refined by using an energy minimisation algorithm (Brooks et al. 1983). The atoms are assigned velocities from a Maxwellian distribution at a temperature below the desired temperature. The system is then equilibrated by integrating the equation of motion while adjusting the temperature and density to the appropriate values. In the definition of temperature one has to include in this context the total number of unconstrained degrees of freedom $(3N - n)$; it is therefore given by

$$T(t) = \frac{1}{(3N - n)k_B} \sum_{i=1}^{N} m_i \mathbf{v}_i^2. \tag{6.16}$$

The equilibration time of such a system takes up to 75 ps simulation time.

Before MD simulations were used to model such biomolecules one assumed this structure to be basically static. However, these calculations showed that the polymer atoms are in constant motion at ordinary temperatures. A typical protein performs a variety of motions driven by collisions either of the protein with the solvent molecules, which can lead to irregular elastic deformations of the entire protein, or collisions with neighbouring atoms in the protein which can result in the liberation of interior groups (for an overview of protein motions see Table 6.1). Accordingly, the dynamics cover a wide range with amplitudes between 0.01 and 100 Å, and energies from 0.1 to 100 kcal/mol on timescales of 10^{-15} to 10^3 s.

The first simulations of biomolecules studied individual molecules composed of around 500 atoms that had potential functions which could be constructed sufficiently accurately to give meaningful results of the approximate equilibrium properties (McCammon et al. 1977). These early molecular dynamics applications were limited to molecules in vacuo; that is, in the absence of any solvent, and the timespan of the simulation was a few hundred picoseconds. However, these simulations provided key insight into the importance of flexibility in biological functions. With standard MD methods it is now possible to

Table 6.1. *Internal motions of globular proteins*

I. Local motions:	0.01 to 5 Å	10^{-15} to 10^{-1} s
(a) Atomic fluctuations		
(b) Sidechain motions		
(c) Loop motions		
(d) Terminal arm motions		
II. Rigid-body motions	1 to 10 Å	10^{-9} to 1 s
(a) Helix motions		
(b) Domains (hinge-bending) motions		
(c) Subunit motions		
III. Larger scale motions:	over 5 Å	10^{-7} to 10^{4} s
(a) Helix coil transitions		
(b) Dissociation/association and coupled structural changes		
(c) Opening and distortional fluctuations		
(d) Folding and unfolding transitions		

study biomolecules in solution and simulate up to nanoseconds. It is clear from studies in molecular fluids, where simulations of several nanoseconds exist, that corresponding extensions of the timescale will also be necessary in the area of macromolecules (Swope & Anderson 1984).

This extension of the MD method could be provided by the hierarchical tree algorithm by speeding up the Coulombic part of the force calculation. Of all the terms in the potential and the force, the Coulombic term typically takes by far the longest time to compute. Although some implementations truncate the range of the Coulombic field to reduce the computational effort (Brooks et al. 1983), this is thought to give bad energy conservation (Windemuth 1991). Another way of reducing the computational costs is to update the forces of nearby particle pairs more frequently than those of more distant particle pairs. This can be done as described in Chapter 4 because the forces of distant particle pairs change less rapidly than those of near neighbours (Grubmuller 1991). Apart from that, alternative methods like harmonic dynamics, stochastic dynamics, and activated dynamic methods are used (for a detailed description see Brooks (1988)).

To our knowledge, octagonal hierarchical tree codes have not been used until now for biomolecule simulation, but the FMM algorithm by Greengard and Roklin (1987) has been applied to model biological macromolecules.

Board et al. (1992) showed that the calculation time of a 24,000-atom system was reduced by an order of magnitude with a small sacrifice in accuracy; with higher accuracy the calculation time of the direct particle–particle method was still more than triple that of the FMM code. Shimada, Kaneko, and Takada (1994) showed that FMM codes can reduce the error in the potential felt by each particle to 0.1–1 kcal/mol, which is much smaller than the 30 kcal/mol involved in conventional methods using truncations in order to reduce computation time.

In these FMM implementations, two modifications to standard PP codes are usually made: The FMM code includes the calculation of the van der Waals forces only for the atom pairs for which the interaction is calculated directly, but not in the interactions evaluated by multipole and Taylor expansions. Therefore, the van der Waals forces are truncated unsystematically, but at distances where the forces have almost vanished due to the $1/r^6$ dependence. The first modification concerns the chemically bonded atoms. They are usually excluded from the calculation of nonbond forces, which is difficult to incorporate in an FMM algorithm. So the excluded forces are calculated separately by a standard MD code and then subtracted from the full force calculated by FMM, which makes the results less accurate.

Unless biomolecules in vacuum are modelled or reaction field or related methods are applied, protein simulations utilise periodic boundary conditions. The central cell is generally cubic and we saw in the previous chapter how such boundary conditions can be implemented in tree codes. However, in some cases the central cell is chosen to have a truncated octahedral or even a more general geometry (van Gunsteren & Berendson 1984, Theodoron & Suter 1985). These kind of problems could not be treated by a tree algorithm in its present form. Apart from this very special case, hierarchical tree codes clearly have enormous potential in the simulation of large inhomogeneous macromolecular systems, especially if one includes different timesteps for near neighbours and distant particles.

6.4 Materials Science

The development of materials for modern technology applications demands a fundamental understanding of the microstructure and kinetic processes underlying a variety of macroscopic behaviours. In materials science one is interested in the behaviour of solids with defects and the effect of extreme environmental conditions like high pressure, temperature, or external stress. The properties of crystals are strongly influenced by the presence of defects such as surfaces, grain boundaries, dislocations, and inclusions.

A problem of longstanding interest is the finite temperature behaviour of solids whose properties are controlled by processes occurring at or near grain boundaries; for example, in developing materials for fusion energy systems and for semiconductor devices carrying high currents, it is necessary to understand how temperature and stress affect the local structure and the kinetics at interfaces. It would be difficult or impossible to study the temperature dependence of the thermodynamic quantities which are defect-sensitive by traditional solid-state physics techniques. MD and MC simulations have the advantage in this context that they are able to study all equilibrium and nonequilibrium properties at the atomic level without the restriction to low temperatures, long times, or small amplitudes. They can even handle strongly nonlinear processes like melting (Abraham 1983), crystal growth (Gilmer 1980), diffusion kinetics (Murch & Rothman 1983, Jacucci & McDonald 1975), and crack propagation.

The use of periodic boundaries for a simulation model containing a crystal defect may generate additional defect structure, which in turn could appreciably influence the system's behaviour. Wang and Lesar (1995) have developed an FFM code to study dislocation dynamics. They include long-range interactions between parallel edge and screw dislocations. They calculate the stress terms from the complex potentials of the material and expand those terms into multipole series.

7

The Fast Multipole Method

In previous chapters we occasionally referred to an alternative type of tree code, namely the Fast Multipole Method (FMM). This technique, an elegant refinement of the basic Barnes–Hut algorithm, appears to be best suited to 'static' problems, where the particle distribution is more or less uniform. Although it has not been as widely used as the Barnes–Hut (BH) method for dynamic problems – because of either its increased mathematical complexity or the additional computational overhead – it may well become the basis of 'multimillion' N-body problems in the near future. We therefore include an introduction to FMM here, based primarily on works by Greengard (1987, 1988, 1990) and Schmidt and Lee (1991). At the same time, we will try to maintain a consistency of notation with the Barnes–Hut algorithm (hereafter referred to as the 'tree method' or 'tree algorithm'), as described in Chapter 2.

7.1 Outline of the Fast Multipole Algorithm

The Fast Multipole Method makes use of the fact that a multipole expansion to infinite order contains the total information of a particle distribution. As in the BH algorithm, the interaction between near neighbours is calculated by direct particle–particle force summation, and more distant particles are treated separately. However, the distinction between these two contributions is obtained in a different way. In FMM the distant region is treated as a *single* 'far-field' contribution, which is calculated by a high-order multipole expansion.

The FMM was first formulated by Greengard and Rokhlin (1987). By forming an infinite multipole expansion on the lowest level of the tree and carefully combining and shifting these centres up and down the tree, the N-body problem can be reduced to an order $O(N)$ algorithm. Of course, the multipole expansion cannot be performed to infinity, but arbitrary accuracy (e.g., within numerical

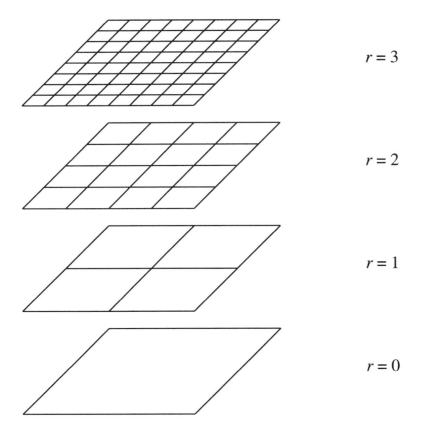

$r = 3$

$r = 2$

$r = 1$

$r = 0$

Fig. 7.1. Division of the simulation box in a fast multipole code for a 2D example.

rounding error) can be assured a priori by taking a given number of terms in the expansion.

Like the tree algorithm, FMM starts with a box big enough to contain all the simulation particles, and this box is subsequently subdivided into boxes of length $d/2^r$ ($r = 0, 1, 2, \ldots$) equivalent to 8^r equal sized subvolumes (4^r in two dimensions). In contrast to the tree method, however, this is done for *every single box* up to a given maximum refinement level R, regardless of the number of particles it contains (see Fig. 7.1). This maximum refinement level R is chosen so that the number of boxes is approximately equal to the number of the simulated particles N. This means that at refinement level R there is an arbitrary number of particles per box, but assuming now that the N particles are more or less homogeneously distributed, there will be *on average* one particle in each box on refinement level R. To fulfill this condition the maximum refinement

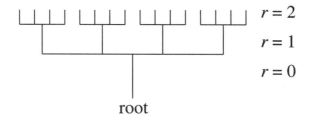

Fig. 7.2. Tree structure of a fast multipole code.

level is chosen as

$$
\begin{aligned}
R &= \log_4 N, && \text{in 2D,} \\
R &= \log_8 N, && \text{in 3D.}
\end{aligned}
\tag{7.1}
$$

The division procedure is used to build a hierarchical structure such as the one shown for a 2D example in Fig. 7.2. The FMM tree structure is used to distinguish different refinement levels r and to decide whether boxes are 'near neighbours'. For every box i at refinement level r, near neighbours are defined as the box itself and any box at the same level with which it shares a boundary point. Interaction lists for each box are later used to ensure that the multipole expansions are shifted appropriately up and down the tree. By definition, a box on the same level which is *not* in a near-neighbour list is well separated: A local multipole expansion made about the centre of this box will then automatically be valid.

At the highest refinement level R a multipole expansion is performed for every single box, for example, for box A in Fig. 7.3. Although the infinite multipole expansion would contain the total information about the system, this obviously has to be truncated at some point. In fact, it is possible to determine the number of terms needed in the expansion to compute the potentials and/or forces of the interactions to a given accuracy ϵ. Usually, the maximum number of terms L in the multipole expansion is chosen such that (Greengard & Rokhlin 1987, Ambrosiano et al. 1988):

$$
\left(\sum_i |q_i| \right) 2^{-L} \leq \epsilon.
\tag{7.2}
$$

Because the average number of particles per box on the refinement level R is 1, this reduces to

$$
2^{-L} \leq \epsilon.
\tag{7.3}
$$

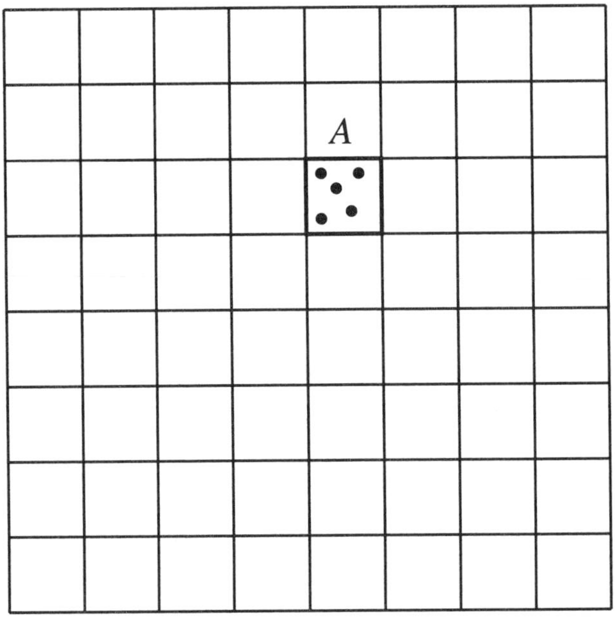

Fig. 7.3. Multipole expansion is performed box-by-box starting at the finest level.

Figure 7.4 illustrates the required refinement level for a given number of particles, and Fig. 7.5 shows the number of terms necessary in the expansion to achieve a certain accuracy. In contrast to the tree method, the multipole expansion of each box is not calculated relative to the centre of mass of the particles, but to the *centre of the box*. In the next step the multipole moments on the next coarser refinement level $r = R - 1$ are calculated. Just as for the tree method, the shifted multipole moments of the daughter cells can be used to obtain the multipole expansion of the parent cell on this lower (i.e., next coarser) refinement level – see Section 2.3. Here again, the moments are calculated relative to the centre of the cell (see Fig. 7.6). Eventually, the moment expansions are carried down to refinement level $r = 0$, which then contains an L-term expansion for the whole system. We note in passing that for consistency with Chapter 2, we use the opposite nomenclature to that used by many authors in the literature, whose trees tend to be upside-down, that is, one descends from the root to the deepest refinement levels! Throughout this book, however, trees are *ascended* from the root up to the *highest* refinement level.

Having come down the hierarchical structure to obtain the multipole moments we now go up again. In this upward pass, a distinction is made in the interaction

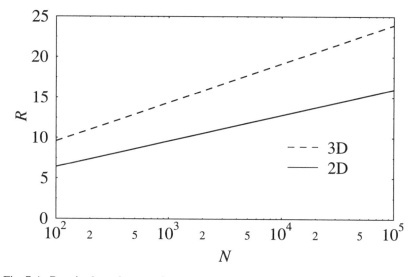

Fig. 7.4. Required maximum refinement level R as a function of the number of simulation particles N according to Eq. 7.1 for 2- and 3-dimensional problems.

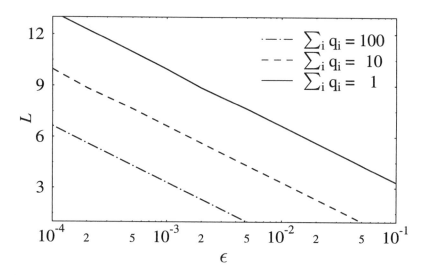

Fig. 7.5. Number of terms L necessary in the multipole expansion to achieve a certain accuracy ϵ with an average of 1, 10, and 100 particles per box at a maximum refinement level R.

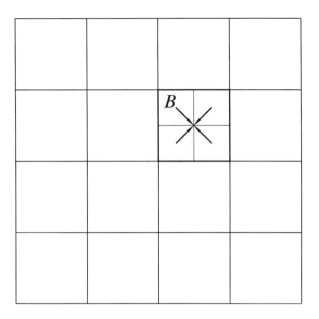

Fig. 7.6. Shifting of the multipole expansion from the daughters to the parents during the downward pass.

list between particle–particle and particle-far-field. There are actually *three* regions: the near field, the 'interactive' field, and the far field. The near field consists of the neighbouring cells; the far field is the entire simulation box excluding the cell in question and its neighbours. The interactive field is the part of the far field that is contained in the near field of this cell's *parents* (see Fig. 7.7).

In the upward pass, each multipole expansion is converted into a local expansion (i.e., a Taylor expansion about the centre of all well-separated boxes at each level). The region of validity of the multipole expansions that contribute to the local expansion consists of all boxes at the same level which are not near neighbours. At levels $r = 0$ and $r = 1$, no boxes are distant enough that a conversion of multipole to local expansion could be made. From $r = 2$ to $r = R$ the following operations can be performed.

For each box on level r the local expansion of the parent box is shifted to each centre of the daughters (see Fig. 7.8). In this local expansion of the daughter cell there are now boxes missing. These are the boxes that do not touch the current daughter box, but do not contribute to the local expansion of the parent cell either – in other words, the boxes of the *interactive* field (see Fig. 7.9). Their contribution has to be added to the local expansion of the daughter cell. The

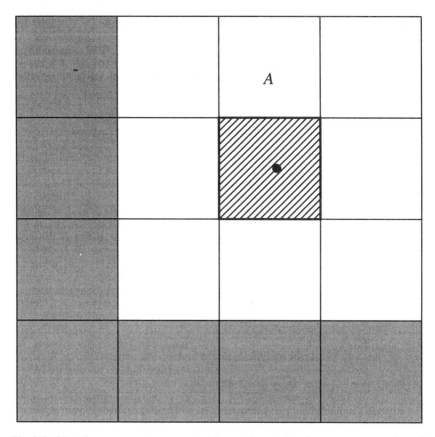

Fig. 7.7. List of well-separated boxes which contribute to local expansion of the shaded box at level $r = 2$.

result is the local expansion due to all particles in all boxes at the same level which are not near neighbours. However, the aim is to evaluate the potential or force not on the box, but on the particles inside the box. Therefore, once the highest refinement level is reached, the local expansions at the individual particle locations have to be evaluated.

Finally, the remaining interactions with the particles in neighbouring boxes and the box itself are added by direct summation of the particle–particle interactions (see Fig. 7.10). Figure 7.11 shows the whole process of the FMM in a flowchart. The mathematical validity of the multipole shifting, the transformation of the multipole in the local expansion, and the shifting of the local

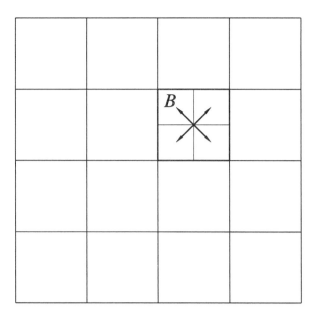

Fig. 7.8. Decomposition of the local expansion from the parents to the daughters during the upward pass.

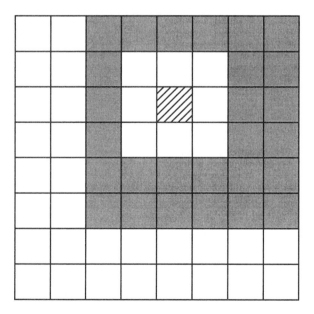

Fig. 7.9. List of well-separated boxes which contribute to the local expansion of the shaded box at level $r = 3$.

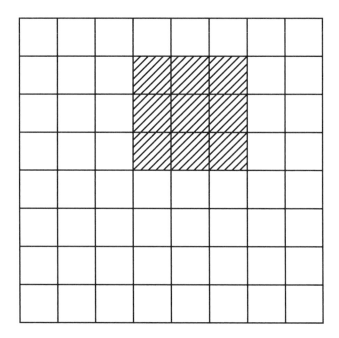

Fig. 7.10. Near-neighbour boxes for the direct particle–particle sum at the finest level.

expansion is discussed in detail by Greengard and Rohklin (1987), Carrier et al. (1988), and Greengard (1990).

How does one actually perform these expansions and shiftings? The first FMM codes (Greengard & Rokhlin 1987, Ambrosiano et al. 1988) were two-dimensional and exploited the complex variable notation to represent the potentials and fields. The 3D calculation is somewhat different, using spherical harmonics instead, so we describe these two cases separately in Sections 7.2 and 7.3. Some recent improvements to these formulations – in which the central multipole transformations are optimised – have been proposed by Petersen et al. (1994) and White and Headgordon (1994).

7.2 2D Expansion

The expressions for the potential Φ and the force $\mathbf{F}$ for gravitational and electrostatic problems in two dimensions are given in Appendix 1. Since a large number of terms (typically 8–20) are needed for the implementation of FMM, it is essential to find a concise mathematical representation of the expansions so that they can be manipulated according to the procedures outlined previously.

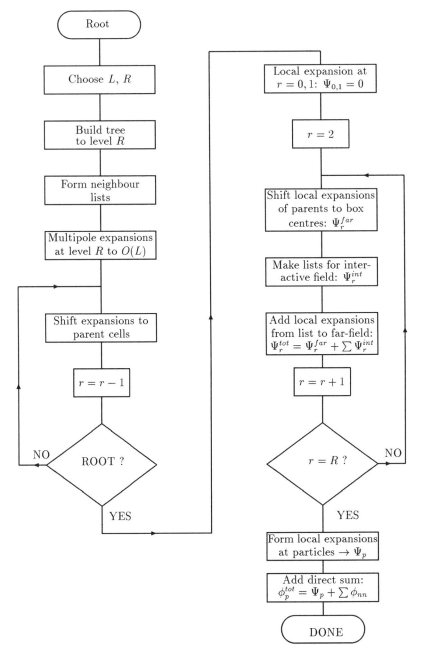

Fig. 7.11. Flowchart of the fast multipole code.

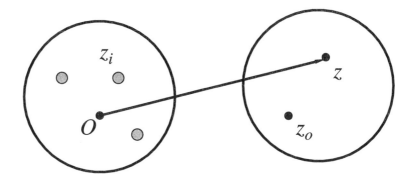

Fig. 7.12. Geometry for the multipole expansion in a 2D complex plane.

Greengard and Rohklin (1987) used the complex plane C to describe a particle's position z_i and charge q_i or mass m_i. The complex potential Φ and the force $\mathbf{F}$ are

$$\Phi(z) = \sum_i q_i \log(z - z_i) \tag{7.4}$$

$$\mathbf{F}(z) = q \sum_i \frac{q_i}{z - z_i}, \tag{7.5}$$

where q is substituted by m for gravitational problems. The *physical* potential is then given by:

$$\Phi(x, y) = Re\{\Phi(z)\},$$

and the force field by:

$$\mathbf{E}(x, y) = \left(Re\{\Phi'(z)\}, \ -Im\{\Phi'(z)\}\right). \tag{7.6}$$

Now consider the situation in Fig. 7.12. We wish to find an expansion for the potential valid in region B due to an arbitrary distribution of charges $q_i(z_i)$ in region A. Rewriting (7.4) and expanding in z_i/z, we have:

$$\begin{aligned}
\Phi(z) &= \sum_{i=1}^{m} q_i \left[\log z + \log\left(1 - \frac{z_i}{z}\right)\right] \\
&= \sum_{i=1}^{m} q_i \log z - \sum_{i=1}^{m} \frac{q_i z_i}{z} - \sum_{i=1}^{m} \frac{q_i z_i^2}{2z^2} + \cdots \\
&= \sum_{i=1}^{m} q_i \log z - \sum_{k=1}^{\infty} \sum_{i=1}^{m} q_i \frac{q_i z_i^k}{k z^k}. \tag{7.7}
\end{aligned}$$

This expression is valid for $|\frac{z_i}{z}| < 1$, that is, for any z which is well separated from the multipole cluster. The error ξ in truncating the k-sum after s terms is

$$\xi \leq \left(\frac{\sum_{i=1}^{m} |q_i|}{\frac{z}{d} - 1} \right) \left(\frac{z}{d} \right)^s, \tag{7.8}$$

where d is the system size. This means that to obtain a certain accuracy, ξ has to be smaller than the required relative precision ϵ. Therefore, from (7.2), the number of terms in the expansion has to be approximately $- \log_{(z/d)} \epsilon$.

In the next step the multipole expansion has to be shifted. Suppose that the multipole expansion due to a set of charges with the centre at z_o is given by

$$\Phi(z) = a_o \log(z - z_o) + \sum_{k=1}^{\infty} \frac{a_k}{(z - z_o)^k}, \tag{7.9}$$

where $a_o = \sum_{i=1}^{m} q_i$, and $a_k = - \sum_{i=1}^{m} (q_i z_i^k)/(k)$. Then the multipole expansion at the same point z, centred at the *origin* is given by:

$$\Phi(z) = a_o \log(z) + \sum_{l=1}^{\infty} \frac{b_l}{z^l},$$

where

$$b_l = - \sum_{i=1}^{m} \frac{q_i (z_i + z_o)^l}{l}$$

$$= - \sum_{i=1}^{m} \frac{q_i}{l} \left[z_o^l + \sum_{k=1}^{l} \binom{l}{k} z_i^k z_o^{l-k} \right]$$

$$= - \left[\frac{a_o z_o}{l} + \sum_{k=1}^{l} \frac{k}{l} \binom{l}{k} \frac{q_i z_i^k}{k} z_o^{l-k} \right]$$

$$= - \frac{a_o z_o^l}{l} + \sum_{k=1}^{l} \binom{l-1}{k-1} a_k z_o^{l-k}, \tag{7.10}$$

where $\binom{l}{k}$ are the binomial coefficients and a_k, $k = 0, 1, \ldots, l$ are defined as before. Equation 7.10 can now be used to calculate the multipole expansion according to the centre of the cell and to go down the tree to evaluate the multipole expansion of the parent cells.

The transformation of the multipole expansion into the Taylor series (local expansion) can be obtained by expanding (7.9) about z_o. Rearranging, we have:

$$\Psi(z) = a_o \log(-z_o) + a_o \log\left(1 - \frac{z}{z_o}\right)$$

$$+ \sum_{k=1}^{\infty} \frac{(-1)^k}{z_o^k} a_k \left(1 - \frac{z}{z_o}\right)^{-k}$$

$$= a_o \log(-z_o) - \sum_{k=1}^{\infty} \frac{1}{k} \left(\frac{z}{z_o}\right)^k + B,$$

$$(7.11)$$

where

$$B = \sum_{k=1}^{\infty} (-1)^k \frac{a_k}{z_o^k} \left[1 + \sum_{p=1}^{\infty} (-1)^p \frac{k(k+1)\ldots(p+k-1)}{p!} \frac{z^p}{z_o^p}\right].$$

The sums over p and k can be re-ordered to give:

$$B = \sum_{k=1}^{\infty} (-1)^k \frac{a_k}{z_o^k} + \sum_{l=1}^{\infty} z^l \left[\frac{1}{z_o^l} \sum_{k=1}^{\infty} (-1)^k \frac{a_k}{z_o^k} \binom{l+k-1}{k-1}\right].$$

In obtaining this expression, we have made use of the identity:

$$\binom{n}{k} = \binom{n}{n-k}.$$

Thus, we have for the local expansion:

$$\Psi(z) = \sum_{l=0}^{\infty} b_l z^l, \qquad (7.12)$$

where

$$b_o = a_o \log(-z_o) + \sum_{k=1}^{\infty} (-1)^k \frac{a_k}{z_o^k},$$

and

$$b_l = -\frac{a_o}{l z_o^l} + \frac{1}{z_o^l} \sum_{k=1}^{\infty} (-1)^k \frac{a_k}{z_o^k} \binom{l+k-1}{k-1}, \quad l \geq 1. \qquad (7.13)$$

For going up the tree again, translations of the local expansion have to be performed. This can be done by making use of the identity:

$$\sum_{k=0}^{n} b_k (z - z_o)^k = \sum_{l=0}^{n} c_l z^l,$$

where

$$c_l = \sum_{k=l}^{n} b_k \begin{pmatrix} k \\ l \end{pmatrix} (-z_o)^{k-l}. \tag{7.14}$$

In this 2-dimensional description we followed closely the works by Greengard. For more mathematical details of the region of validity and the error bounds see Greengard (1987) and Ambrosiano et al. (1988).

7.3 3D Expansion

The 3D FMM was also formulated by Greengard (1988), but Schmidt and Lee (1991) described a practical implementation with some numerical tests. Using the mathematical conventions for spherical harmonics and units as in Jackson (1975), the multipole expansion of the potential is given by

$$\Phi(\mathbf{r}) = 4\pi \sum_{l,m} \frac{M_{lm} Y_{lm}(\theta, \phi)}{(2l+1)r^{l+1}} \tag{7.15}$$

in spherical coordinates (r, θ, ϕ) relative to an origin O, with the multipole moments:

$$M_{lm} = \sum_{i} q_i s_i^l Y_{lm}^*(\theta_i, \phi_i), \tag{7.16}$$

where the charges have the coordinates (s_i, θ_i, ϕ_i). The connection between this representation (to lth order) and the Cartesian geometry used in Chapters 2 and 5 for the multipole expansion to quadrupole order is described in Appendix 2.

Now shifting the origin O to O' by a translation vector $\mathbf{r}_t$ a transformation of the multipole moments is needed to obtain the expansion in terms of the new vector $\mathbf{r}' = \mathbf{r} - \mathbf{r_t}$ relative to O'. This transformation is given by:

$$M'_{l'm'} = \sum_{l,m} T^{MM}_{l'm',lm} M_{lm},$$

with the transformation matrix $T^{MM}_{l'm',lm}$ being

$$T^{MM}_{l'm',lm} = 4\pi \frac{(-r_t)^{l'-l} Y^*_{l'-l,m'-m}(\theta_t, \phi_t) a'_{l'-l,m'-m} a_{lm} (2l'+1)}{2(l+1)[2(l'-l)+1] a_{l'm'}},$$

and a_{lm} is defined as

$$a_{lm} = (-1)^{l+m} \frac{2(l+1)^{1/2}}{[4n(l+m)!(l-m)!]^{1/2}}.$$

Using these equations the shifting can be used to obtain the multipole moments of the parent cells, just as described in Section 2.3. The local expansion in this notation is given by:

$$\Psi(\mathbf{r}) = 4\pi \sum_{l,m} L_{lm} r^l Y_{lm}(\theta, \phi),$$

where L_{lm} are referred to as the local moments of the Taylor series expansion. The local moments are obtained from the multipole moments by the transformation:

$$L'_{l'm'} = \sum_{l,m} T^{LM}_{l'm',lm} M_{lm}, \tag{7.17}$$

where the transformation matrix $T^{LM}_{l'm',lm}$ is

$$T^{LM}_{l'm',lm} = 4\pi \frac{(-1)^{l+m} Y^*_{l'+l,m'-m}(\theta_t, \phi_t) a_{l,m} a_{l'm'}}{r_t^{l'+l+1}(2l+1)(2l'+1) a_{l'+l,m'-m}}. \tag{7.18}$$

Finally, we have to calculate the shifting when going down the tree. The shifted local moments of the daughter cells can be obtained by

$$L'_{l'm'} = \sum_{l,m} T^{LL}_{l'm',lm} L_{lm},$$

with

$$T^{LL}_{l'm',lm} = 4\pi \frac{r_t^{l-l'} Y_{l'-l,m'-m}(\theta_t, \phi_t) a_{l'm'} a_{l-l',m-m'}}{(2l'+1)[2(l-l')+1] a_{lm}}.$$

Fortunately, the transformation matrices are not dense. The region of validity and the error bounds are discussed by Greengard (1988) and Schmidt and Lee (1991).

7.4 Implementation of Fast Multipole Codes

The implementation of the tree-building routine in FMM is actually easier than for the usual tree code because boxes are always divided into four or eight subcells, and all boxes on a given level are equivalent: There is no need to distinguish between 'leaves' and 'twigs' – there are only twig nodes. One way of doing the subdivision (in 2D) is shown next. To facilitate comparison, the variable names and notation correspond to that used earlier in Chapter 2, Table 2.1, except that the nodes are labelled with positive integers (0, 1, 2, 3 ...).

```
pointer = 2
inode = 1
node(1) = 0
iparent = 1
idau(1) = 1

do r = 0, R – 1
    nbox = 4^r
    for each parent at level r, create 4 daughters:
    do i = 1, nbox
        idau(iparent) = inode
        do j = 1, 4
            parent(pointer) = node(iparent)
            node(pointer) = inode
            level(inode) = r
            r_c(inode) = r_c(iparent) + r_s(j)/2^r
            inode = inode + 1
            pointer = pointer + 1
        end do
        iparent = iparent + 1
    end do
end do
```

The shift vector $\mathbf{r}_s$ has the values $(1, 1)$, $(1, -1)$, $(-1, 1)$, $(-1, -1)$ for $j = 1, \ldots, 4$, respectively. The tree structure for $R = 2$ is shown in Table 7.1.

As long as the nodes are labelled according to this scheme, the pointer array 'ipoint' is somewhat redundant. We have kept it, however, to illustrate the contrast between FMM and the recursive structure of the BH tree. Moreover, it is also a fairly straightforward matter to adapt the tree-build algorithm described in Chapter 2. For instance, in the flowchart of Fig. 2.4, we replace the decision box 'Is cell a twig?' with: 'Is the current node undivided AND below the maximum refinement level?'. If so, we divide further; if not, we check its sister node or return to its parent node. The advantage of retaining the original tree infrastructure is that it is easily extended to the adaptive version of FMM (Carrier et al. 1988).

At each step in the potential or force calculation, it is necessary to know which boxes are near neighbours, and which contribute to the near-field interaction list. One possible algorithm to obtain these lists is as follows.

Table 7.1. *Tree arrays corresponding to Fig. 7.2*

pointer	level	node	parent	1st_dau
1	0	0	0	2
2	1	1	0	6
3	1	2	0	10
4	1	3	0	14
5	1	4	0	18
6	2	5	1	–
7	2	6	1	–
8	2	7	1	–
9	2	8	1	–
10	2	9	2	–
11	2	10	2	–
12	2	11	2	–
13	2	12	2	–
14	2	13	3	–
15	2	14	3	–
16	2	15	3	–
17	2	16	3	–
18	2	17	4	–
19	2	18	4	–
20	2	19	4	–
21	2	20	4	–

do r = 1, R
    nbox = 4^r
    set pointer to 1st box on level r

    do i = 1, nbox
        inode = node(pointer)
        iparent = parent(pointer)

        loop over parent's interaction list

        do k = 1, nlist(iparent)
            nn-parent = ilist(iparent,k)
            for each daughter j of nn-parent **do**
                jnode = node(j)

> **if** inode and jnode touching **then**
> update near-neighbour list for inode
> **else**
> update interaction list for inode
> **endif**
> **end do**
> **end do**
>
> **end do**
> **end do**

This routine looks longer than it is. In fact, at the highest refinement level, it requires an effort $O(36 \times 4^R)$, compared with $O(4^{2R})$ if we did the searches by examining all box-pairs at this level.

Evaluation of the potential and forces can now proceed according to Fig. 7.11 using the expansions and shifting formulae shown previously. More arrays are needed in an FMM code than in a standard tree code to store the intermediate local expansions for the 'far' and 'interaction' fields corresponding to each box. These will typically be 2-dimensional arrays, for example, Ψ^{far}(inode, l), containing L local expansion coefficients (7.13, 7.17) summed over the far field of box inode.

Various boundary conditions can be implemented by the appropriate choice of the local expansions at levels $r = 0$ and $r = 1$. For example, fully periodic boundary conditions can be included efficiently by adding the periodic images to the interaction list of the $r = 1$ cells in the 'upward pass' (root-to-leaves) of the algorithm (see Fig. 7.11) (Greengard & Rokhlin 1987, Greengard 1988, Schmidt & Lee 1991).

7.5 Timing and Accuracy

Having seen how FMM codes work in principle, we now consider the conditions under which we can expect an $O(N)$ scaling of computational effort. At first sight it seems that FMM codes with a computation time proportional to N are superior to hierarchical tree codes with a $N \log N$ and particle–particle codes with an N^2 dependence of the computation time. Moreover, FMM codes have the additional advantage of giving an upper error bound for the calculation. However, FMM codes have quite a big overhead due to the multipole and Taylor expansions, so one has to ask instead: At which number of particles is it appropriate to use an FMM code instead of a PP or tree code? This is not a simple question to answer, because the computation time depends not only on the number of particles N, but also on how accurately the calculation should be performed.

For the tree code, accuracy is determined by the tolerance parameter, θ; for the FMM code, the number of multipole terms, L, and the refinement level, R, have the same function. Schmidt and Lee (1991) showed that the overall polynomial dependence of the computation on N, L, and R is given by

$$P = aBL^2 + bNL^2 + cBL^4 + dB\left[g_1 + g_2\left(\frac{N}{B}\right) + g_3\left(\frac{N}{B}\right)^2\right],$$

$$\text{(7.19)}$$

where a, b, c, g_1, g_2, and g_3 are machine-dependent parameters and B is the number of boxes on the highest refinement level ($B = 4^R$ in 2D and $B = 8^R$ in 3D). Equation 7.18 shows that it is crucial that the number of boxes B equal to the number of simulation particles, because only then is the N^2 dependence removed, and the FMM algorithm becomes $O(N)$.

We have seen that although the structure of the FMM algorithm is the same in two and three dimensions, its actual implementation is rather different. The first suggested fast multipole algorithm (Greengard & Rokhlin 1987) was a two-dimensional version with $B = 4^R$ and L chosen according to Eq. 7.3. Figure 7.13 shows a comparison between this FMM code and a particle–particle code (Greengard 1988).

Figure 7.14 shows a comparison of the computation time required by a particle–particle code, tree code, and fast multipole codes for the 3-dimensional case.

In some of these 2- and 3-dimensional examples L and R have not been selected according to Eqs. 7.1 and 7.3, but are just chosen as free parameters. However, free choice of L and R can involve some problems, because a smaller R is equivalent to a less refined grid, which means that N/B, the average number of particles per box, becomes larger. Keeping a certain accuracy has two consequences – according to Eq. 7.3 the number of terms L in the multipole expansion has to increase, enlarging the computational overhead, and according to Eq. 7.18 the interaction of particles in neighbouring boxes takes N^2 time. This term therefore determines when it is useful to go to a higher maximum refinement level R as N increases. But additional refinement levels contribute significantly to the computational costs, therefore it is necessary to have the quadratic term coefficient as small as possible to avoid this. So choosing L and R freely can be justified as long as the problem does not require high accuracy

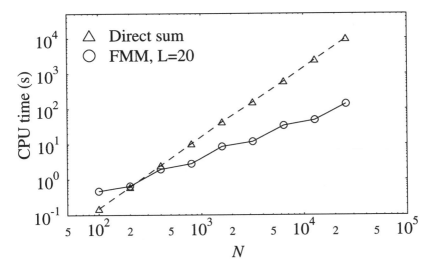

Fig. 7.13. Comparison of computation time as a function of the number of simulation particles N between particle–particle and the fast multipole code for 2-dimensional uniform charge distribution.

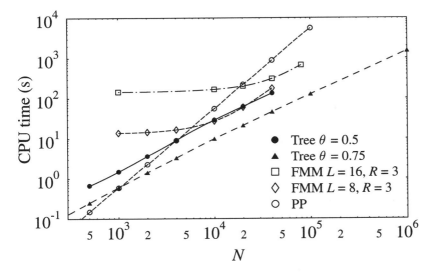

Fig. 7.14. Comparison of computation time as a function of the number of simulation particles N between particle–particle, hierarchical tree, and the fast multipole code for 3-dimensional problems. The FMM timings are taken from Schmidt and Lee (1991).

of the calculation and some of the requirements can be dropped, but this should not be done in general.

One of the assumptions of the described FMM code was that the distribution of the particles is more or less homogeneous. Nonuniform particle distributions would require either a high refinement level or a large number of terms in the multipole expansion, both resulting in high computational costs. Therefore, the original form of the fast multipole method is not very suitable for nonuniform distributions, or for dynamical systems which may develop large density contrasts over the course of time. For 2-dimensional problems a so-called adaptive fast multipole code (Carrier et al. 1988) has been developed, which is basically a hybrid between the tree and the fast multipole algorithms. It divides the physical space like the tree code, but instead of stopping the division process when there is just one particle in the box the division ends when the number of particles is smaller than a given number s. Apart from this division process the force calculation is performed as in the fast multipole algorithm, with s determining the number of terms necessary in the multipole expansion.

7.6 Applications

Two-dimensional fast multipole codes have been used to calculate such different problems as the dynamics of an electron beam emitted near a grounded plate (Ambrosiano et al. 1988) and the nonbond forces in dynamical protein systems (Board Jr. et al. 1992). Not only can such open boundary problems be treated, but also problems with periodic, Dirichlet, and Neumann conditions (Greengard & Rohklin 1987, Ambrosiano et al. 1988, Engheta et al. 1992).

Three-dimensional implementations of the fast multipole algorithm have been suggested by Schmidt (1991) and some numerical tests have been performed by Carrier et al. (1988) and Zhao (1991).

Recently, FMM codes have been applied in many different fields. In most cases they are not used just to follow the evolution of N particles, but are combined with other methods. In the following we will mention some examples of such applications. This list is in no way complete, but serves to emphasize the broad range of applications already using FMM.

In incompressible fluids dynamics (Chorin & Marsden 1990) the flow may be modelled by the dynamics of a set of N interacting vortices. Their interaction is described by the Biot–Savart equation

$$\mathbf{B}(\mathbf{x}) = \frac{1}{c} \int \mathbf{J}(\mathbf{y}) \times \frac{\mathbf{x} - \mathbf{y}}{|\mathbf{x} - \mathbf{y}|^3} d\mathbf{y}.$$

Although vortices are objects with a vector character, mathematically their interaction is very similar to gravitational or electrostatic problems. Salmon et al. (1994) developed a fast parallel tree code which can be applied either to gravitational applications or to fluid dynamical problems. Schmitt (1994) used an FMM method to apply contour dynamics to evolve regions of constant vorticity moving under Eulerian equations. Strickland (1994) presented examples for complicated geometries like disks, disks with holes, hemispheres, and hemispheres particularly with applications for parachutes. Additional work on FMM for vortex problems has been done by Russo and Strain (1994).

The fast multipole method can be used to solve electromagnetic scattering problems from 3-dimensional conducting bodies of arbitrary shape. Murphy, Rokhlin, and Vassiliou (1993), Song and Chew (1994), Hoekstra and Sloot (1994), and Wagner and Chew (1994) showed how to discretise the electric field integral equation into moments and how to use an FMM to compute the matrix-vector multiplications.

One very difficult problem is suspensions, for which it is necessary to include unit cells with many close-to-touching inclusions. These calculations require a high accuracy. Helsing (1994b) showed how using an FMM code in combination with the conjugated gradient method allows unit cells with thousands of inclusions to be treated with very high accuracy. Similar calculations were performed for two-component composites (Helsing 1994a).

Kuhn and Muller (1993) have suggested using hierarchical tree codes and FMM in the context of high-level concepts represented by agents, environments, and controllers based on and supported by the object-oriented paradigm. They demonstrate the flexibility of such simulation systems for applications related to environmental research and air quality control.

There has been a lot of discussion in the literature about whether the Barnes–Hut algorithm or the Fast Multipole Method is ultimately 'better' for solving N-body problems. We should emphasise that this is not an either/or question: It all depends on the type of problem at hand. It is worth noting, however, that because most of the effort in FMM is spent in computing the far-field multipole expansions for every *box* at the highest refinement level, it is not clear that the method could be combined effectively with individual timesteps (as described in Chapter 4). The BH algorithm has a clear advantage here because the effort in determining interaction lists depends directly on the number of particles for which the force is evaluated. As a rule, it appears that standard tree codes are more suited to dynamic problems where $N \sim 10^4$–10^5 and accuracy requirements are not so stringent, whereas FMM codes are more effective for static problems where accuracy is important and $N > 10^6$. This is by no means the end of the story, and one can expect to see hybrid codes emerging in the near future with adaptive tree structures, limited multipole expansions, and well-defined precision.

Appendix 1

Multipole Expansion in Two Dimensions

There are many physical situations that do not need the full three-dimensional treatment described in the previous chapters, and a one- or two-dimensional description is often more appropriate to describe the essential physics of the problem. Assuming that the problem can be simplified with a suitable geometry, using a lower dimension is equivalent to higher numbers of simulation particles. Seen another way, for the same number of particles per unit length, area, or volume, one gets better statistics in 1D and 2D than in 3D.

Having in mind gravitational (or electrostatic $m \equiv q$) problems with a $1/r$-potential in three dimensions the force law between two particles in each dimension is given by:

$$1D : \mathbf{F} = -2\pi m \frac{\mathbf{x}}{|\mathbf{x}|},$$

$$2D : \mathbf{F} = -2m \frac{\mathbf{r}}{|\mathbf{r}|^2},$$

$$3D : \mathbf{F} = -m \frac{\mathbf{r}}{|\mathbf{r}|^3}.$$

To appreciate the physical origin of the 1D and 2D force laws (actually the field as just written), it is helpful to visualise the particles as mass sheets and mass rods, respectively. In electrostatic problems one can think of capacitor plates and line charges.

The potentials corresponding to these forces are:

$$1D : \Phi(x) = 2\pi m |\mathbf{x}|,$$

$$2D : \Phi(\mathbf{r}) = 2m \log |\mathbf{r}|,$$

$$3D : \Phi(\mathbf{r}) = -\frac{m}{|\mathbf{r}|}.$$

In Chapter 2 we saw that the multipole expansion is an essential ingredient for reducing the computation time of the force calculation with a hierarchical tree code. Here we calculate this expansion for the two-dimensional case. As in the three-dimensional case the potential at the origin Φ due to the pseudoparticles is the sum of the potentials Φ_i due to the particles in the cell, but now with the two-dimensional potential given previously:

$$\Phi(\mathbf{R}) = \sum_i \Phi_i(\mathbf{R} - \mathbf{r}_i) = 2\sum_i m_i \log |\mathbf{R} - \mathbf{r}_i|, \tag{1.1}$$

where $\mathbf{R}$ and $\mathbf{r}_i$ are defined as in Fig. 2.10. The multipole expansion is given by

$$\Phi(\mathbf{R}) = 2\sum_i m_i \left[1 - \mathbf{r}_i \frac{\partial}{\partial \mathbf{r}} + \frac{1}{2}\mathbf{r}_i\mathbf{r}_i : \frac{\partial}{\partial \mathbf{r}}\frac{\partial}{\partial \mathbf{r}} + \dots \right] \log R.$$

There are in principle three kinds of derivatives:

$$\frac{\partial}{\partial x} \log R = \frac{x}{R^2},$$

$$\frac{\partial}{\partial x}\frac{\partial}{\partial x} \log R = \frac{1}{R^2} - \frac{2x^2}{R^4}$$

$$\frac{\partial}{\partial x}\frac{\partial}{\partial y} \log R = -\frac{2xy}{R^4}.$$

Obtaining the other coordinates in the same manner, the multipole expansion of the potential due to the pseudoparticle is

$$\Phi(\mathbf{R}) = 2\sum_i m_i \left[\log R - x_i \frac{x}{R^2} - y_i \frac{y}{R^2} \right.$$

$$+ \frac{1}{2}x_i^2 \left(\frac{1}{R^2} - \frac{2x^2}{R^4} \right)$$

$$\left. + \frac{1}{2}y_i^2 \left(\frac{1}{R^2} - \frac{2y^2}{R^4} \right) - x_i y_i \left(\frac{2xy}{R^4} \right) \right].$$

The force between the particle and the pseudoparticle is given as the derivative of the potential $\mathbf{F} = -m_p \frac{\partial}{\partial \mathbf{r}}\Phi_i$. The x-component essentially consists of the following terms.

Monopole: $\dfrac{\partial}{\partial x} \log R = \dfrac{x}{R^2}.$

Dipole: $\dfrac{\partial}{\partial x} \left(x_i \dfrac{x}{R^2} + y_i \dfrac{y}{R^2} \right) = \left(\dfrac{1}{R^2} - \dfrac{2x^2}{R^4} \right) x_i - \dfrac{2xy}{R^4} y_i.$

Quadrupole: $\dfrac{\partial}{\partial x} \left[\dfrac{1}{2} x_i^2 \left(\dfrac{1}{R^2} - \dfrac{2x^2}{R^4} \right) + \dfrac{1}{2} y_i^2 \left(\dfrac{1}{R^2} - \dfrac{2y^2}{R^4} \right) - x_i y_i \left(\dfrac{2xy}{R^4} \right) \right]$

$$= \dfrac{1}{2} x_i^2 \left(\dfrac{8x^3}{R^6} - \dfrac{6x}{R^4} \right) + \dfrac{1}{2} y_i^2 \left(\dfrac{8xy^2}{R^6} - \dfrac{2x}{R^4} \right)$$

$$- x_i y_i \left(\dfrac{2y}{R^4} - \dfrac{8x^2 y}{R^6} \right).$$

The components of the force vector in two dimensions are therefore given by:

$$F_x^p = -2m_p \left[\dfrac{x}{R^2} \cdot \sum_i m_i + \left(\dfrac{2x}{R^4} - \dfrac{1}{R^2} \right) \cdot \sum_i m_i x_i + \dfrac{2xy}{R^4} \cdot \sum_i m_i y_i \right.$$

$$+ \left(\dfrac{8x^3}{R^6} - \dfrac{6x}{R^4} \right) \cdot \dfrac{1}{2} \sum_i m_i x_i^2$$

$$+ \left(\dfrac{8xy^2}{R^6} - \dfrac{2x}{R^4} \right) \cdot \dfrac{1}{2} \sum_i m_i y_i^2$$

$$\left. + \left(\dfrac{8x^2 y}{R^6} - \dfrac{2y}{R^4} \right) \cdot \sum_i m_i x_i y_i \right].$$

$$F_y^p = -2m_p \left[\dfrac{y}{R^2} \cdot \sum_i m_i + \left(\dfrac{2y}{R^4} - \dfrac{1}{R^2} \right) \cdot \sum_i m_i y_i + \dfrac{2xy}{R^4} \cdot \sum_i m_i x_i \right.$$

$$+ \left(\dfrac{8y^3}{R^6} - \dfrac{6y}{R^4} \right) \cdot \dfrac{1}{2} \sum_i m_i y_i^2$$

$$+ \left(\dfrac{8x^2 y}{R^6} - \dfrac{2y}{R^4} \right) \cdot \dfrac{1}{2} \sum_i m_i x_i^2$$

$$\left. + \left(\dfrac{8xy^2}{R^6} - \dfrac{2x}{R^4} \right) \cdot \sum_i m_i x_i y_i \right].$$

The shifting of the centres of mass works exactly as described in Chapter 2.

Appendix 2
Spherical Harmonics

As we saw in Section 7.2, spherical harmonics allow the 3D multipole expansion to be expressed in a compact manner and to be manipulated as required by the fast multipole algorithm. Here we provide a few more details for readers reluctant to reach for their copy of Jackson, and to provide a link with the Cartesian geometry used in Chapters 2 and 5.

The multipole expansion in spherical coordinates (r, θ, ϕ) can be written as (Jackson 1975):

$$\Phi(\mathbf{r}) = \sum_{l=0}^{\infty} \sum_{m=-l}^{l} \frac{4\pi}{2l+1} q_{lm} \frac{Y_{lm}(\theta, \phi)}{r^{l+1}}, \tag{2.1}$$

where

$$q_{lm} = \sum_i q_i s_i^l Y_{lm}^*(\theta_i, \phi_i), \tag{2.2}$$

and Y_{lm} are the spherical harmonic functions, defined next. The charges are assumed to have positions s_i, such that $z_i = s \cos\theta$, $x = s \sin\theta \cos\phi$, and $y = s \sin\theta \sin\phi$. The first few spherical harmonics up to the second order (i.e., $l = 2$) are as follows.

$$Y_{00} = \frac{1}{\sqrt{4\pi}},$$

$$Y_{10} = \sqrt{\frac{3}{4\pi}} \cos\theta,$$

$$Y_{11} = -\sqrt{\frac{3}{8\pi}} \sin\theta e^{i\phi},$$

$$Y_{20} = \sqrt{\frac{5}{4\pi}} \left(\frac{3}{2} \cos^2\theta - \frac{1}{2} \right),$$

$$Y_{21} = -\sqrt{\frac{15}{8\pi}} \sin\theta \cos\theta e^{i\phi},$$

$$Y_{22} = \frac{1}{4}\sqrt{\frac{15}{2\pi}} \sin^2\theta e^{2i\phi}. \tag{2.3}$$

Additional terms can be found from:

$$Y_{lm}(\theta, \phi) = \sqrt{\frac{2l+1}{4\pi}\frac{(l-m)!}{(l+m)!}} P_l^m(\cos\theta)e^{im\phi},$$

where

$$P_l^m(x) = (-1)^m (1-x^2)^{m/2} \frac{d^{m+l}}{dx^{m+l}}(x^2-1)^l$$

are the Legendre polynomials. The expressions in Eq. 2.2 are related to the Cartesian multipole moments as follows.

$$q_{00} = \frac{M}{\sqrt{4\pi}}, \tag{2.4}$$

$$q_{10} = \sqrt{\frac{3}{4\pi}} D_z, \tag{2.5}$$

$$q_{11} = -\sqrt{\frac{3}{8\pi}}(D_x - iD_y), \tag{2.6}$$

$$q_{20} = \frac{1}{2}\sqrt{\frac{5}{4\pi}} Q_{zz}, \tag{2.7}$$

$$q_{21} = -\frac{1}{3}\sqrt{\frac{15}{8\pi}}(Q_{xz} - iQ_{yz}), \tag{2.8}$$

$$q_{22} = \frac{1}{12}\sqrt{\frac{15}{2\pi}}(Q_{xx} - Q_{yy} - 2iQ_{xy}). \tag{2.9}$$

To recover the potentials in Cartesian coordinates, we substitute (2.9) and (2.3) in (2.1) to get:

$$\Phi_0 = 4\pi q_{00}\frac{Y_{00}}{r}$$

$$= \frac{M}{r},$$

$$\Phi_1 = \sum_{m=-1}^{1} \frac{4\pi}{3} q_{1m} \frac{Y_{1m}}{r^2}$$

$$= \sum_j \frac{r_j D_j}{r^3},$$

$$\Phi_2 = \sum_{jk} \frac{1}{2} Q_{jk} \frac{r_j r_k}{r^5}, \qquad (2.10)$$

where

$$M = \sum q,$$

$$D_j = \sum q s_j,$$

$$Q_{jk} = \sum q (3 s_j s_k - s^2 \delta_{jk})$$

are the monopole, dipole, and quadrupole moments, respectively. The sum over charges is implied. The expressions for the potential in Eq. 2.10 correspond to those in Eq. 2.7 in Chapter 2.

Appendix 3
Near-Neighbour Search

The hierarchical tree method can not be adapted only for Monte Carlo applications: It can also be modified to perform near-neighbour searches efficiently. This means that the tree algorithm could also have applications for systems with short-range or contact forces. Hernquist and Katz (1989) first showed how the tree structure can be used to find near neighbours through range searching. Following their method, the near-neighbour search is performed the following way.

Consider a system in which only neighbours lying within a distance h will interact with the particle i in question. For the near-neighbour search this sphere is enclosed in a cube whose sides are of length $2h$. The tree is built the usual way and the tree search starts at the root. The tree search is performed in a very similar way to the normal force calculation of Section 2.2 by constructing an interaction list. The main difference is that the s/d criterion is substituted by the question: 'Does the volume of the search cube overlap with the volume of the pseudoparticle presently under consideration?'

If there is no overlap, this branch of the tree contains no near neighbours and is not searched any further. However, if there is an overlap, the cell is subdivided into its daughter cells and the search continues on the next highest level. If the cell is a leaf – meaning there is only one particle in the cell – one has to check whether it actually lies within the radius h of particle i. In this case it will be added to the interaction list. This has to be done for all paths in the tree.

Hernquist and Katz performed the tree search by a node-by-node procedure, but a vectorised level-by-level search provides no additional difficulties. In fact, the interaction list algorithm described in Section 4.3 can be readily adapted along the lines just outlined.

For large particle numbers the near-neighbour search using the tree algorithm is obviously more efficient than methods that compare particle–particle distances. However, one should mention that there are grid-based methods (see, for example, Hockney and Eastwood (1981)), which can be up to a factor 3–5 faster than this tree-based method (Benz 1988). Usually the search grid Δ is chosen so that $\Delta = h$, since for $\Delta < h$ the performance of these codes decreases.

On the other hand, these methods have the disadvantage that the cell width has to be the same for all particles. The tree method can handle different search-area sizes without difficulties and is therefore much more flexible. For this reason tree codes have already been recognised as a promising means of speeding up Smooth Particle Hydrodynamic (SPH) codes.

References

Aarseth, S. J. 1963. Dynamical evolution of clusters of galaxies. I. *Mon. Not. Roy. Astro. Soc.*, **126**, 223–255.

Aarseth, S. J., and Hoyle, F. 1964. An assessment of the present state of the gravitational N-body problem. *Astrophysica Norwegica*, **29**, 313–321.

Aarseth, S. J., Lin, D. N., and Palmer, P. L. 1993. Evolution of planetesimals. II. Numerical simulations. *Astrophys. J.*, **403**, 351–376.

Abraham, D. B. 1983. Surface reconstruction in crystals. *Phys. Rev. Lett*, **51**, 1279–1281.

Adams, D. J. 1975. Grand canonical ensemble Monte Carlo for a Lennard–Jones fluid. *Mol. Phys.*, **29**, 307–311.

Ahmad, A., and Cohen, L. 1973. A numerical integration scheme for the N-body gravitational problem. *J. Comp. Phys.*, **12**, 389–402.

Aichelin, J., and Stöcker, H. 1986. Quantum molecular dynamics – a novel approach to N-body correlations in heavy ion collisions. *Phys. Lett. B*, **176**, 14.

Alder, B. J., and Wainwright, T. E. 1959. Studies in molecular dynamics. I. General method. *J. Chem. Phys.*, **31**, 459–466.

Allen, M. P., and Tildesley, D. J. 1987. *Computer simulations of liquids*. Oxford: Oxford University Press.

Alper, H. E., Bassolino, D., and Stouch, T. R. 1993. Computer simulation of a phospholipid monolayer-water system: The influence of long-range forces on water-structure and dynamics. *J. Chem. Phys.*, **98**, 9798–9807.

Ambrosiano, J., Greengard, L., and Rokhlin, V. 1988. The fast multipole method for gridless particle simulation. *Comp. Phys. Commun.*, **48**, 117–125.

Appel, A. 1985. An efficient program for many-body simulation. *SIAM J. Sci. Statist. Comput.*, **6**, 85.

Applegate, J. H., Douglas, M. R., Gürsel, Y., Hunter, P., Seitz, C. L., and Sussman, G. J. 1985. A digital orrery. *IEEE Trans. Comput. Sci.*, **C-34**, 86–95.

Barnes, J., and Hernquist, L. 1993. Computer models of colliding galaxies. *Physics Today*, March, 54–61.

Barnes, J., Hernquist, L., and Schweizer, F. 1991. Colliding galaxies. *Scientific American*, 40–47.

Barnes, J., and Hut, P. 1986. A hierarchical $O(N \log N)$ force-calculation algorithm. *Nature*, **324**, 446–449.

Barnes, J. E. 1990. A modified tree code: don't laugh; it runs. *J. Comp. Phys.*, **87**, 161–170.

Baus, M. 1977. Computer models for colliding galaxies. *Physica A*, **88**, 319.

Beeman, D. 1976. Some multistep methods for use in molecular dynamics calculations. *J. Comp. Phys.*, **20**, 130.

Belhadj, M., Alper, H. E., and Levy, R. M. 1991. Molecular-dynamics simulations of water with Ewald summation for the long-range electrostatic interactions. *Chem. Phys. Lett.*, **179**, 13–20.

Benz, W. 1988. Applications of smooth particle hydrodynamics (SPH) to astrophysical problems. *Comp. Phys. Commun.*, **48**, 97–105.

Bernu, B. 1981. One-component plasma in a strong uniform magnetic field. *J. de Physique.*, **42**, L253–L255.

Beveridge, D. L., Mezei, M., Mehrotra, P. K., Marchese, F. T., Ravi-Shankar, G., Vasu, T., and Swaminathan, S. 1983. Monte-Carlo computer simulation studies of the equilibrium properties and structure of liquid water. Pages 297–351 of: *Advances in Chemistry*. Washington, D.C.: American Chemical Society.

Binder, K. 1987. *Applications of Monte Carlo methods in statistical physics*. Heidelberg: Springer.

Birdsall, C.K., and Langdon, A.B. 1985. *Plasma physics via computer simulation*. New York: McGraw-Hill.

Board Jr., J. A., Causey, J. W., Leathrum Jr., J. F., Windemuth, A., and Schulten, K. 1992. Accelerated molecular dynamics simulation with the parallel fast multipole algorithm. *Chem. Phys. Lett.*, **198**, 89–94.

Boersch, H. 1954. Experimentelle Bestimmung der Energieverteilung in thermisch ausgelösten Elektronenstrahlen. *Z. Physik*, **139**, 115–146.

Borchardt, I., Karantzoulis, E., Mais, H., and Ripken, G. 1988. Calculation of beam envelopes in storage rings and transport systems in the presence of transverse space charge effects and coupling. *Z. Phys. C*, **39**, 339–349.

Bouchet, F. R., and Hernquist, L. 1988. Cosmological simulations using the hierarchical tree method. *Astrophys. J. Suppl.*, **68**, 521–538.

Bouchet, F. R., and Hernquist, L. 1992. Gravity and count probabilities in an expanding universe. *Astrophys. J.*, **400**, 25–40.

Brackbill, J. U., and Ruppel, H. M. 1986. FLIP – a method for adaptively zoned, particle-in-cell calculations of fluid-flows in 2 dimensions. *J. Comp. Phys.*, **65**, 314–343.

Brooks, C. L., III. 1987. The influence of long-range force truncation on the thermodynamics of aqueous ionic-solutions. *J. Chem. Phys.*, **86**, 5156–5162.

Brooks, C. L., III, Bruccoleri, R., Olafson, B., Swaninathan, S., and Karplus, M. 1983. *J. Comp. Chem.*, **4**, 187.

Brooks, C. L., III, Karplus, M., and Pettit, B. M. 1988. *Proteins: A theoretical perspective of dynamics, structure and thermodynamics*. New York: John Wiley.

Brush, S. G., Sahlin, H. L., and Teller, E. 1966. Monte Carlo study of a one-component plasma. I. *J. Chem. Phys.*, **45**, 2102–2118.

Caillol, J. M. 1992. Search of the gas-liquid transition of dipolar hard-spheres. *J. Chem. Phys.*, **96**, 7039–7053.

Carrier, J., Greengard, L., and Rokhlin, V. 1988. A fast adaptive multipole algorithm for particle simulations. *SIAM J. Sci. Stat. Comput.*, **9**, 669–686.

Cassing, W., and Mosel, U. 1990. Many-body theory of high-energy heavy-ion reactions. *Prog. Part. Nucl. Phys.*, **25**, 235–323.

Ceperley, D. M., and Alder, B. J. 1987. Ground-state of solid hydrogen at high-pressures. *Phys. Rev. B*, **36**, 2092–2106.

Chabrier, G., Ashcroft, N. W., and DeWitt, H. E. 1992. White-dwarfs as quantum crystals. *Nature*, **360**, 48–50.

Chao, A. W. 1983. Coherent instabilities of a relativistic bunched beam. Page 353 of: Month, M. (ed), *Physics of high energy particle accelerators*. New York: AIP.

Chen, F. F. 1974. *Introduction to plasma physics*. New York: Plenum Press.

Chorin, A. J., and Marsden, J. E. 1990. *A mathematical introduction to fluid mechanics*. New York: Springer.

Ciccotti, G. 1991. Computer simulations of equilibrium and nonequilibrium molecular dynamics. Pages 943–969 of: Hansen, J.P., Levesque, D., and Zinn-Justin, J. (eds.), *Liquids, freezing and glass transition*. Amsterdam: Elsevier Science Publishers.

Clementi, E., and Chakravorty, S. J. 1990. A comparative study of density functional models to estimate molecular atomization energies. *J. Chem. Phys.*, **93**, 2591–2602.

Csernai, L. P. 1994. *Introduction to relativistic heavy ion collisions*. Chichester: John Wiley.

Dawson, J. M. 1983. Particle simulation of plasmas. *Rev. Mod. Phys.*, **55**, 403–410.

Deutsch, C. 1977. Nodal expansion in a real matter plasma. *Phys. Lett. A* **60**, 317–318.

Dharma-wardana, M. W. C. 1988. In: F. J. Rogers and H. E. Dewitt, *Strongly coupled plasma physics*. New York: Plenum Press.

Dubinski, J., and Carlberg, R. G. 1991. The structure of cold dark matter halos. *Astrophys. J.*, **378**, 496.

Eastwood, J., and Hockney, R. W. 1974. Shaping the force law in two-dimensional particle-mesh models. *J. Comp. Phys.*, **16**, 342.

Efstathiou, G., Davis, M., Frenk, C. S., and White, S. D. M. 1985. Numerical techniques for large cosmological *N*-body simulations. *Astrophys. J. Suppl.*, **57**, 241–260.

Elson, R., Hut, P., and Inagaki, S. A. 1987. Dynamical evolution of globular clusters. *Rev. Astro. Astrophys.*, **25**, 565–601.

Engheta, N., Murphy, W. D., Rokhlin, V., and Vassiliou, M. S. 1992. The fast multipole method (FMM) for electromagnetic scattering problems. *IEEE Trans. Antennas and Prop.*, **40**, 634–641.

Ewald, P. P. 1921. Die Berechnung optischer und elektrostatischer Gitterpotentiale. *Ann. Physik*, **64**, 253.

Fawley, W. M., Laslett, L. J., Celata, C. M., Faltens, A., and Haber, I. 1993. Simulation studies of space-charge-dominated beam transport in large-aperture ratio quadrupoles. *Il Nuovo Cim.*, **106**, 1637.

Friedman, A., Callahan, D. A., Grote, D. P., Haber, I., Langdon, A. B., and Lund, S. M. 1993. What we have learned from 3D and *r,z* intense-beam simulation using the WARP code. *Il Nuovo Cim.*, **106**, 1649–1655.

Friedman, A., Grote, D. P., Callahan, D. A., and Langdon, A. B. 1992. 3D particle simulation of beams using the WARP code: transport around bends. *Part. Acc.*, **37–38**, 131–139.

Gibbon, P. 1992a. TREEQMD: A hierarchical tree code for quantum molecular dynamics simulations. *IBM Technical Report*, **TR 75.92.21**.

Gibbon, P. 1992b. The vectorised tree code: A new work horse for *N*-body problems. *IBM Technical Report*, **TR 75.92.12**.

Gilmer, G.H. 1980. Computer models of crystal growth. *Science*, **208**, 355.

Goldreich, P., and Ward, G. R. 1973. The formation of planetesimals. *Astrophys. J.*, **183**, 1051.

Greenberg, R., Wacker, J. F., Hartmann, W. K., and Chapman, C. R. 1978. Planetesimals to planets: numerical simulation of collisional evolution. *Icarus*, **35**, 1.

Greengard, L. 1988. *The rapid evaluation of potential fields in particle systems*. Cambridge, Mass.: MIT Press.

Greengard, L. 1990. The numerical solution of the *N*-body problem. *Computers in Physics*, Mar./Apr., 142–152.

Greengard, L., and Rokhlin, V. 1987. A fast algorithm for particle simulations. *J. Comp. Phys.*, **73**, 325–348.

Grubmuller, H. 1991. *Mol. Simul.*, **6**, 121.

Haber, I., Kehne, D., Reiser, M., and Rudd, H. 1991. Experimental, theoretical and numerical investigation of the homogenization of density nonuniformities in the periodic transport of a space-charge dominated beam. *Phys. Rev. A*, **44**, 5194–5205.

Hansen, J. P. 1985. Computer simulation of basic plasma phenomena. Pages 433–496 of: *Scottish Summer School Proceedings*.

Hansen, J. P., and McDonald, I. R. 1981. Microscopic simulation of a strongly coupled hydrogen plasma. *Phys. Rev. A*, **23**, 2041–2059.

Hansen, J. P., McDonald, I. R., and Viellefosse, P. 1979. Statistical mechanics of dense ionized matter. VII. Dynamical properties of binary ionic mixtures. *Phys. Rev. A*, **20**, 2590–2602.

Hasse, R. W. 1991. Excess energy of cylindrical Coulomb crystals. *Phys. Rev. Lett.*, **67**, 600–602.

Hayli, A. 1967. The *N*-body problem in an external field. Application to the dynamic evolution of open clusters. I. *Bull. Astron.*, **21**, 67–90.

Helsing, J. 1994a. Bounds on the shear modulus of composites by interface integral methods. *J. Mech. Phys. Solids*, **42**, 1123–1138.

Helsing, J. 1994b. Fast and accurate calculation of structural parameters for suspensions. *Proc. Roy. Soc. A*, **445**, 127–140.

Hernquist, L. 1987. Performance characteristics of tree codes. *Astrophys. J. Supp.*, **64**, 715–734.

Hernquist, L. 1988. Hierarchical *N*-body methods. *Comp. Phys. Commun.*, **48**, 107–115.

Hernquist, L. 1990. Vectorization of tree traversals. *J. Comp. Phys.*, **87**, 137–147.

Hernquist, L. 1993a. *N*-body realization of compound galaxies. *Astrophys. J. Suppl.*, **86**, 389–400.

Hernquist, L. 1993b. Structure of merger remnants. II. progenitors with rotating bulges. *Astrophys. J.*, **409**, 548–562.

Hernquist, L., and Barnes, J. E. 1990. Are some *N*-body algorithms intrinsically less collisional than others? *Astrophys. J.*, **349**, 562–569.

Hernquist, L., Bouchet, F. R., and Suto, Y. 1991. Application of the Ewald method to cosmological *N*-body simulations. *Astrophys. J. Suppl.*, **75**, 231–240.

Hernquist, L., and Katz, N. 1989. TREESPH: A unification of SPH with the hierarchical tree method. *Astrophys. J. Suppl.*, **70**, 419–446.

Hernquist, L., and Mihos, C. 1995. Excitation of activity in galaxies by minor mergers. *Astrophys. J.*, **448**, 41–63.

Hernquist, L., and Quinn, P. 1988. Formation of shell galaxies. *Astrophys. J.*, **331**, 682.

Hockney, R. W., and Eastwood, J. W. 1981. *Computer simulation using particles*. New York: McGraw-Hill.

Hoekstra, A. G., and Sloot, P. M. A. 1994. New computational techniques to simulate light-scattering from arbitrary particles. *Part. Syst. Charact.*, **11**, 189–193.

Hohl, D., Natoli, V., Ceperley, D.M., and Martin, R.M. 1993. Molecular-dynamics in dense hydrogen. *Phys. Rev. Lett.*, **71**, 541–544.

Hut, P., Makino, J., and McMillan, S. L. 1988. Modeling the evolution of globular star-clusters. *Nature*, **336**, 31–35.

Ichimaru, S. 1982. Strongly coupled plasmas: High-density classical plasmas and degenerate electron liquids. *Rev. Mod. Phys.*, **54**, 1017–1059.

Iyetomi, H., and Ichimaru, S. 1986. Theory of interparticle correlations in dense, high-temperature plasmas. 8. Shear viscosity. *Phys. Rev. A*, **34**, 3203–3209.

Jackson, J. D. 1975. *Classical electrodynamics*. New York: John Wiley.

Jackson, M. 1986. Ph.D. thesis, Harvard University, Cambridge, Mass.

Jacucci, G., and McDonald, I. R. 1975. Structure and diffusion in mixtures of rare gas liquids. *Physica*, **80A**, 607–625.

Jansen, G. H. 1990. *Coulomb interaction in particle beams*. Boston: Academic Press.

Jernighan, J. G. 1985. Direct N-body simulations with a recursive center of mass reduction and regularization. *I. A. U. Symp.*, **113**, 275–284.

Jernighan, J. G., and Porter, D. H. 1989. A tree code with logarithmic reduction of force terms, hierarchical regularization of all variables, and explicit accuracy controls. *Astrophys. J. Suppl.*, **71**, 871–893.

Jin, Wei, Rino, J. P., Vashishta, P., Kalia, R. K., and Nakano, A. 1993. Structural and dynamical correlations in glass. Pages 357–366 of: Horn, H. M., and Ichimaru, S. (eds.), *Strongly coupled plasma physics*. Rochester, N.Y: University of Rochester Press.

Jorgensen, W. L., Binning, R. C., and Bigot, B. 1981. Structures and properties of organic liquids. *J. Am. Chem. Soc.*, **103**, 4393.

Kapchinskii, I. M., and Vladimirskii, V. V. 1959. Limitations of proton beam current in a strong focusing linear accelerator associated with the beam space charge. Page 274 of: *Proceedings of the International Conference on High Energy Accelerators*. Geneva: CERN.

Klein, U. 1983. Coexistance of superconductivity and ferromagnetism in surface-layers of rare-earth ternary compounds. *Phys. Lett.*, **96**, 409.

Konopka, J. 1992. Der Einfluß der expliziten Form der Wellenfunktion auf die Dynamik einer Schwerionenkollision. Diploma Thesis, Inst. for Theoretical Physics, Univ. Frankfurt, Germany.

Kruer, W. L. 1988. *The physics of laser plasma interactions*. New York: Addison-Wesley.

Kuhn, V., and Muller, W. 1993. Molecular-dynamics simulations of hot, dense hydrogen. *J. Visualization Comput. Animation*, **4**, 95–111.

Kustannheimo, P., and Stiefel, E. L. 1965. *J. Reine Angew. Math.*, **218**, 204.

Kwon, I., Collins, L. A., Kress, J. D., Troullier, N., and Lynch, D. L. 1994. Molecular-dynamics simulations of hot, dense hydrogen. *Phys. Rev. E*, **49**, R4771–R4774.

Lang, A., Babovsky, H., Cassing, W., Mosel, U., Reusch, H-G., and Weber, K. 1992. A new treatment of Boltzmann-like collision integrals in nuclear kinetic equations. *J. Comp. Phys.*, **106**, 391–396.

Lawson, J. D. 1988. *The physics of charged-particle beams*. 2nd ed. Oxford: Oxford University Press.

Lee, E. P., and Cooper, R. K. 1976. General envelope equation for cylindrically symmetric charged-particle beams. *Part. Acc.*, **7**, 83–95.

Levitt, M. 1969. *J. Mol. Biol.*, **46**, 269.

Li, X-P., and Sessler, A. M. 1994. Crystalline beam in a storage ring: how long can it last? Pages 1379–1381 of: *Proceedings of European Particle Accelerator Conference*.

Lindl, J. D., McCrory, R. L., and Campbell, E. M. 1992. Progress toward ignition and burn propagation in inertial confinement fusion. *Phys. Today*, **45**, 32.

McCammon, J. A., Gelin, B. R., and Karplus, M. 1977. Dynamics of folded proteins. *Nature*, **267**, 585.

McMillan, S. L. W., and Aarseth, S. J. 1993. An $O(N \log N)$ integration scheme for collisional stellar systems. *Astrophys. J.*, **414**, 200–212.

McQuarrie, D. A. 1976. *Statistical mechanics*. New York: Harper and Row.

Makino, J. 1990a. *Comp. Phys.*, **88**, 393.

Makino, J. 1990b. Vectorization of a treecode. *J. Comp. Phys.*, **87**, 148–160.

Metropolis, N., Rosenbluth, A. W., Rosenbluth, M. N., Teller, A. H., and Teller, E. 1953. Equation of state calculations by fast computing machines. *J. Chem. Phys.*, **21**, 1087–1092.

Murch, G. E., and Rothman, S. J. 1983. Application of the Monte-Carlo method to solid-state diffusion. *J. Metals*, **35**, 42.

Murphy, W. D., Rokhlin, V., and Vassiliou, M. S. 1993. Acceleration methods for the iterative solution of electromagnetic scattering problems. *Radio Sci.*, **28**, 1–12.

Murthy, N. S., and Worthington, C. R. 1991. X-ray diffraction evidence for the presence of discrete water layers on the surface of membranes. *Biochem. et Biophys. Acta*, **1062**, 172–176.

Nakagawa, Y., Sekiya, M., and Hayashi, C. 1986. Settling and growth of dust particles in a laminar phase of a low-mass solar nebula. *Icarus*, **67**, 375–390.

Neise, L., Berenguer, M., Hartnack, C., Peilert, G., Stöcker, H., and Greiner, W. 1990. Quantum molecular dynamics – a model for nucleus-nucleus collisions from medium to high energies. *Nuc. Phys. A*, **519**, 375–394.

Pangali, M. R., Rao, M., and Berne, B. J. 1980. A Monte Carlo study of the structure and thermodynamics properties of water: dependence on the system size and on the boundary conditions. *Mol. Phys.*, **40**, 661.

Perry, M. D., and Mourou, G. 1994. Terawatt to petawatt subpicosecond lasers. *Science*, **264**, 917–924.

Peterson, H. G., Soelvason, D., Perram, J. W., and Smith, E. R. 1994. The very fast multipole method. *J. Chem. Phys.*, **101**, 8870–8876.

Pfalzner, S., and Gibbon, P. 1992. An $N \log N$ code for dense plasma simulation. Page 45 of: W. Ebeling, A. Förster, and Radtke, R. (eds.), *Physics of non-ideal plasmas*. Leipzig: Teubner.

Pfalzner, S., and Gibbon, P. 1994. A hierarchical tree code for dense plasma simulation. *Comp. Phys. Commun.*, **79**, 24–38.

Porter, D. 1985. Ph.D. thesis, Department of Physics, University of California, Berkeley.

Press, W. H. 1986. *Page 184 of:* Hut, P., and McMillan, S. L. W. (eds.), *The use of supercomputers in stellar dynamics*. New York: Springer.

Press, W. H., Flannery, B. P., Teukolsky, S. A., and Vetterling, W. T. 1989. *Numerical recipes: the art of scientific computing*. Cambridge: Cambridge University Press.

Rahman, A. 1964. Correlations in the motion of atoms in liquid argon. *Phys. Rev.*, **136**, A405–411.

Rahman, A., and Schiffer, J. P. 1986. Structure of a one-component plasma in an external field: a molecular-dynamics study of particle arrangement in a heavy-ion storage ring. *Phys. Rev. Lett.*, **57**, 1133–1136.

Reusch, H.-G. 1992. Experiences with the parallelization and vectorization of simulation codes for heavy-ion reactions. *Int. J. Supercomp. Applic. High Perf. Comp.*, **6**, 224–240.

Richardson, D. C. 1993. A new tree code method for simulation of planetesimal dynamics. *Mon. Not. R. Astron. Soc.*, **261**, 396–414.

Russo, G., and Strain, J. A. 1994. Fast triangular vortex method for 2d-Euler equations. *J. Comput. Phys.*, **111**, 291–323.

Safronov, V. S. 1969. *Evolution of the protoplanetary cloud and the formation of the earth and the planets*. Moscow: Nauka Press.

Salmon, J. K., and Warren, M. S. 1994. Skeletons from the treecode closet. *J. Comp. Phys.*, **111**, 136–155.

Salmon, J. K., Warren, M. S., and Winckelmans, G. S. 1994. Fast parallel tree codes for gravitational and fluid dynamical N-body problems. *Int. J. Supercomp. Appl. High Perf. Comp.*, **8**, 129–142.

Sangster, M. J., and Dixon, M. 1976. Interionic potentials in alkali halides and their use in simulations of the molten salts. *Adv. in Phys.*, **25**, 247–342.

Schmidt, K. E., and Lee, M. A. 1991. Implementing the fast multipole method in three dimensions. *J. Stat. Phys.*, **63**, 1223–1235.

Schmitt, H. 1994. Contour dynamics and the fast multipole method. *SIAM J. Scient. Comput.*, **15**, 997–1001.

Schofield, P. 1973. *Comp. Phys. Commun.*, **5**, 17.

Schwarz, F. 1986. Colliding and merging of galaxies. *Science*, **231**, 227.

Sellwood, J. A. 1987. The art of N-body building. *Ann. Rev. Astr. Ap.*, **25**, 151–186.

Shimada, J., Kaneko, H., and Takada, T. 1994. Performance of fast multipole methods for calculating electrostatic interactions in biomacromolecular simulations. *J. Comp. Chemistry*, **15**, 28–43.

Slattery, W. L., Doolen, G. D., and DeWitt, H. E. 1982. N-dependence in the classical one-component plasma Monte-Carlo calculations. *Phys. Rev. A*, 2255–2258.

Song, J. M., and Chew, W. C. 1994. Fast multipole method solution using parametric geometry. *Microwave Opt. Techn. Lett.*, **7**, 706–765.

Spitzer Jr., L. 1987. *Dynamical evolution of globular clusters*. Princeton: Princeton University Press.

Steinhauser, O. 1981a. Reaction field simulation of water. *Mol. Phys.*, **45**, 335–348.

Steinhauser, O. 1981b. Single-particle dynamics of liquid carbon-disulfide. *Chem. Phys. Lett.*, **82**, 153–157.

Stevenson, D. J. 1980. The condensed matter physics of planetary interiors. *J. Phys. Suppl.*, **41**, C2–53.

Stevenson, D. J. 1982. Interiors of the giant planets. *Ann. Rev. Earth Planet. Sci.*, **10**, 257–295.

Stillinger, F. H., and Rahman, A. 1974. *J. Chem. Phys.*, **60**, 1545.

Strickland, J. H. 1994. Prediction method for unsteady axisymmetrical flow over parachutes. *J. Aircraft*, **31**, 637–643.

Sugimoto, D. 1993. A pipeline approach to many-body problems. *Physics World*, November, 32–35.

Sugimoto, D., Chikada, Y., Makino, J., Ito, T., Ebisuzaki, T., and Umemura, M. 1990. A special-purpose computer for gravitational many-body problems. *Nature*, **345**, 33–35.

Sugimoto, D., and Makino, J. 1989. Synchronization instability and merging of binary globular-clusters. *Publ. Astro. Soc. Jap.*, **41**, 1117–1144.

Suginohara, T., Suto, Y., Bouchet, F. R., and Hernquist, L. 1991. Cosmological N-body simulations with a tree code: fluctuations in the linear and nonlinear regimes. *Astrophys. J. Supp.*, **75**, 631–643.

Swope, W. C., and Anderson, H. C. 1984. A molecular-dynamics method for calculating the solubility of gases in liquids and the hydrophobic hydration of inert-gas atoms in aqueous solution. *J. Phys. Chem.*, **88**, 6548–6556.

Theodoron, D. N., and Suter, U. W. 1985. *J. Chem. Phys.*, **82**, 955.

Toomre, A., and Toomre, J. 1972. *Astrophys. J.*, **178**, 623.

Tosi, M. P. 1964. *J. Phys. Chem. Solids*, **25**, 45.

Tremaine, S. 1992. The dynamical evidence for dark matter. *Physics Today*, February.

Turq, P. 1977. *J. Chem. Phys.*, **66**, 3039.

van Albada, T. S. 1986. In: Hut, P., and McMillan, S. (eds.), *The use of supercomputers in stellar dynamics*. Berlin: Springer.

van Gunsteren, W. F., and Berendson, H. J. C. 1984. *J. Mol. Biol.*, **176**, 559.

van Horn, H. M. 1980. White dwarfs. *J. Phys. Colloq.*, **41**, C2–97.

Vashishta, P., Kalia, R. K., Antonio, G. A., and Ebbsjö, I. 1989. Atomic correlations and intermediate-range order in molten and amorphous $GeSe_2$. *Phys. Rev. Lett*, **62**, 1651–1654.

Vashishta, P., Kalia, R. K., Rino, J. P., and Ebbsjö, I. 1990. Interaction potential for SiO_2 – a molecular dynamics study of structural correlations. *Phys. Rev. B*, **41**, 12197–12209.

Verlet, L. 1967. Computer experiments on classical fluids. I. Thermodynamical properties of Lennard–Jones molecules. *Phys. Rev.*, **159**, 98–103.

Villumsen, J. V. 1989. A new hierarchical particle-mesh code for very large scale cosmological N-body simulations. *Astrophys. J. Suppl. Series*, **71**, 407–431.

Wagner, R. L., and Chew, W. C. 1994. A ray-propagation fast multipole algorithm. *Microwave Opt. Techno. Lett.*, **7**, 435–438.

Wang, H. Y., and Lesar, R. 1995. $O(N)$ algorithm for dislocation dynamics. *Phil. Mag. A*, **71**, 149–164.

Wertheim, M. S. 1971. *J. Chem. Phys.*, **155**, 4291.

Wetherill, G. W., and Stewart, G. R. 1989. Accumulation of a swarm of small planetesimals. *Icarus*, **77**, 330–357.

White, C. A., and Headgordon, M. 1994. Derivation and efficient implementation of the fast multipole method. *J. Chem. Phys.*, **101**, 6593–6605.

Wiedemann, H. (ed) 1993. *Particle accelerator physics: basic principles and linear beam dynamics*. Berlin: Springer.

Wielen, R. 1967. Dynamical evolution of star cluster models, I. *Veröff. Astr. Recheninstitut Heidelberg*, **19**, 1–43.

Wielen, R. (ed). (1990). *Dynamics and interaction of galaxies*. Heidelberg: Springer.

Wiener, M. C., King, G. I., and White, S. H. 1991. Structure of a fluid dioleoylphosphatidylcholine bilayer determined by joint refinement of x-ray and neutron-diffraction data. 1. Scaling of neutron data and the distributions of double-bonds and water. *Biophys. J.*, **60**, 568–576.

Windemuth, A. 1991. *Mol. Simul.*, **5**, 353.

Wood, W. W. 1968. *Physics of simple liquids*. Amsterdam: North-Holland.

Younger, S. M. 1992. Many-atom screening effects on diffusion in dense helium. *Phys. Rev. A*, **45**, 8657–8665.

Zallen, R. 1983. *The physics of amorphous solids*. New York: John Wiley.

Zérah, G., Clérouin, J., and Pollock, E. L. 1992. Thomas–Fermi molecular dynamics, linear screening and mean-field theories of plasmas. *Phys. Rev. Lett.*, **69**, 446–449.

Zhao, F., and Johnsson, S. L. 1991. The parallel multipole method on the connection machine. *SIAM J. Sci. Stat. Comput.*, **12**, 1420–1437.

Zhou, F., and Schulten, K. 1995. Molecular-dynamics study of a membrane water interface. *J. Phys. Chem.*, **99**, 2194–2207.

Index